Instructor's Manual for

Microbiology

for the
Health Sciences

Instructor's Manual for

Microbiology

for the
Health Sciences

Fourth Edition

Gwendolyn R. W. Burton, Ph.D.

Professor Emeritus
Department of Science
Front Range Community College
Westminster, Colorado

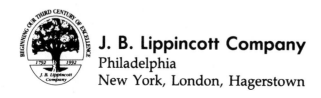

J. B. Lippincott Company
Philadelphia
New York, London, Hagerstown

Sponsoring Editor: Andrew Allen
Cover Designer: Lou Fuiano
Production Supervisor: Robert D. Bartleson
Production Service: Caslon Inc.
Compositor: Digitype, Inc.
Printer/Binder: R. R. Donnelley & Sons
Cover Printer: R. R. Donnelley & Sons

Fourth Edition

6 5 4 3 2 1

ISBN 0-397-54965-2

Introduction

This manual has been prepared to aid instructors who are teaching microbiology classes using the fourth edition of *Microbiology for the Health Sciences*. The students in these classes usually are preparing for careers in nursing, respiratory therapy, dental assisting, radiologic technology, and other health-oriented fields. However, many other students can benefit from the study of the basic principles of microbiology and its relations to health and disease.

The material presented by the instructor may be expanded or reduced to meet the needs of each class. The 10 chapters in the textbook may be used as the basis for a one-credit class for dental assistants, a two-credit class for non-science majors, or a three- or four-credit class for nurses.

The basic approach to teaching microbiology is usually through lectures and laboratory exercises, supplemented by films, discussions, field trips, demonstrations, research papers, and other projects. These learning procedures should be coordinated so that each supplements the material covered in the textbook.

To aid in planning the course, this manual includes a list of objectives for the student, an outline of the material covered in the textbook, a list of sources of audiovisual aids and pertinent films, a summary and discussion of each chapter, suggested laboratory exercises, and multiple-choice test questions. Completion, true-false, and essay questions are found at the end of each chapter in the textbook. It is important to use various types of test questions to evaluate and supplement the students' learning. Frequent quizzes may be given to encourage the students to study the material as it is presented and to detect any misunderstanding or lack of knowledge before the final examination.

The instructor should encourage the students to maximize their learning of this difficult subject by taking advantage of every learning mechanism available to them: (1) studying the text, (2) studying the glossary, (3) answering questions at the end of each chapter, (4) answering self-test questions, (5) taking notes during each lecture, (6) attending each lecture to determine what material will be emphasized and explained further, (7) completing the laboratory work and submitting a laboratory write-up, and (8) viewing the films shown to illustrate the material. Thus, all the students can learn the material in the way they learn best: by repetition, visual images, hearing the discussion, or working in the laboratory.

Contents

Instructor's Manual for

Microbiology

for the
Health Sciences

Outline and Discussion of Chapters

Chapter 1
Introduction to Microbiology

Objectives

1. Define microbiology
2. List some important functions of microbes in the environment
3. Explain the relevance of microbiology to the health professions
4. List some areas of microbiological study
5. Outline some contributions of Leeuwenhoek, Pasteur, and Koch to microbiology
6. Explain the biological theory of fermentation
7. Explain the germ theory of disease
8. Learn Koch's postulates and give some circumstances in which they may not apply
9. Describe the difference between light microscopes and electron microscopes and the applications of both
10. List the metric units used in microscopic measurements and indicate their relative sizes

Summary and Discussion

In Chapter 1 the students are introduced to the significance of microbiology and the microbes in the world around them, especially as they pertain to the health sciences. It is important for them to understand that most microbes are useful and that only a small percentage are pathogenic. The instructor will be able to dwell on many additional illustrations of helpful microorganisms.

Some of the various areas of microbiology are discussed. For health oriented students, the many aspects of medical microbiology are the most interesting, but they should be aware of the many other fields involved with various aspects of microbiology, for example, water and sewage treatment, food and milk processing, agriculture, and industry.

The brief history of microbiology is a good method of introducing the background and fundamentals of microbiology such as epidemiology, immunology, contagion, fermentation, disease causation, and the importance of proper nursing, medical, and surgical techniques.

Outline

I. Microbiology, the science
   A. Microbiology: the study of very small living organisms
      1. Occurrence of microbes
      2. Relevance to health sciences
   B. The scope of microbiology
      1. General microbiology
      2. Medical microbiology
      3. Veterinary microbiology
      4. Agricultural microbiology
         a. Food microbiology
         b. Dairy microbiology
      5. Sanitary microbiology
      6. Industrial and applied microbiology
      7. Microbial genetics
      8. Environmental microbiology
   C. Milestones of microbiology
      1. Discovery of microorganisms
      2. Spontaneous generation
      3. Theory of biogenesis
      4. Biological theory of fermentation
      5. Germ theory of disease
      6. Koch's postulates
      7. Sterilization and disinfection techniques
II. The tools of microbiology
   A. Microscopes
      1. Light microscopy
      2. Electron microscopy
   B. Units of measurement

Suggested Laboratory Exercises

1. Teach the student the proper use and care of the light microscope.
2. Demonstrate the oil immersion technique.
3. Allow the student to examine prepared slides of the three shapes of bacteria and the normal organisms found in the mouth.

Suggested Audiovisual Aids

1. *Importance of Microorganisms.* 16-mm movie, color with sound, 28 min. (MG)
2. *Of Men and Minds.* 16-mm movie, color with sound, 29 min. (SQ)
3. *Microbiology, No. 10, The Germ Theory of Disease (AIBS, Part II).* 16-mm, color with sound, 28 min. (IU)
4. *What is Microbiology?* Videotape EP–2295 (EI)

Sample Examination Questions

1. The study of microbiology is
   a. the study of microscopes
   b. the study of very small insects that are too small to inspect macroscopically
   c. the study of only small species of the animal and plant kingdoms
   *d. the study of microorganisms and their relationships
2. Microbiology is
   a. the study of *only* disease-producing microorganisms
   *b. the study of all biological activities of microbes
   c. used only in the field of medicine
   d. the microscopic study of the human reproductive system
3. Pathogenic organisms are
   *a. disease-producing microorganisms
   b. harmless organisms
   c. indigenous microflora
   d. dead microbes
   e. enzymes
4. Pathogenic microorganisms are
   *a. capable of causing disease
   b. capable of curing disease
   c. organisms that always cause disease
   d. essential to the continuance of our species
   e. none of the above
5. Organisms that have become established in an area of the human body are referred to as
   *a. indigenous microflora
   b. opportunists
   c. symbionts
   d. pathogens
6. The earliest descriptions of microorganisms were made by
   a. Redi
   b. Lister
   c. Pasteur
   d. Mike O. Scope
   *e. Antony van Leeuwenhoek
7. The first man to use a simple magnifying lens to observe microorganisms was
   a. Pasteur
   b. Koch
   *c. Leeuwenhoek
   d. Mixon
   e. Fracastorius
8. The first man to use an antiseptic in surgery to prevent infection was
   a. Kissinger
   *b. Lister
   c. Koch
   d. Pasteur
   e. Tyndall
9. The highest magnification possible with Leeuwenhoek's lenses was about
   a. 100×
   b. 200×
   *c. 300×
   d. 700×
   e. 1000×
10. All of the following are true of Louis Pasteur *except*
    a. he provided the final disproof of spontaneous generation
    b. he was the first to use the terms *aerobic* and *anaerobic*
    *c. he was the first to discover the invisible world of microorganisms
    d. he studied chemistry and crystallography and did his early work in these fields
    e. he devoted most of his time to studies of microorganisms
11. Which of the following men disproved by experimentation the theory of spontaneous generation?
    a. Leeuwenhoek
    b. Lister
    *c. Pasteur
    d. Koch
12. Lister's contribution to microbiology was associated with
    *a. antiseptic surgery
    b. the discovery of Listerine
    c. the design of the Lister bag
    d. preservation of foods
13. The man given credit for the first description of bacteria is

a. Pasteur
*b. Leeuwenhoek
c. Gram
d. Koch

14. The first person to use repeated boiling and cooling to destroy bacterial spores was
    a. Semmelweis
    *b. Tyndall
    c. Pasteur
    d. Koch
    e. Lister

15. What did Koch introduce to the scientific world?
    a. Antiseptic technique
    b. Disproval of the theory of spontaneous generation
    *c. A scientific approach to the field of medical microbiology
    d. The discover of the invisible world of microorganisms
    e. Sterile surgical procedures

16. Robert Koch was responsible for
    a. prevention of infection following surgery by using aseptic technique
    b. the discovery of penicillin
    *c. the postulates to establish a cause-and-effect relationship between a microorganism and a disease
    d. the method of acid-fast staining

17. Koch established
    *a. a scientific approach to studying microorganisms
    b. the difference between procaryotic and eucaryotic cells
    c. the theory of spontaneous generation
    d. that all pathogenic microbes are anaerobic

18. Tyndallization is the process of
    *a. boiling and reboiling to establish a sterile mixture
    b. washing one's lab coat daily
    c. heating an instrument in a flame before it is used
    d. drying wheat
    e. making whiskey

19. Sterilization is synonymous with
    a. disinfection
    b. sanitization
    c. antiseptic technique
    *d. none of these

20. The process of disinfection refers to
    *a. destruction of disease-producing organisms
    b. sterilization
    c. removal of all bacteria
    d. destruction of all bacterial spores

21. Useful magnifications, using the oil immersion objective and the 10✕ ocular, can be obtained to

a. 10✕
b. 100✕
*c. 1,000✕
d. 10,000✕
e. none of the above

22. A micrometer, formerly called a micron, is a unit of measure of microorganisms. It is equal to
    *a. 0.001 mm
    b. 0.01 mm
    c. 0.1 mm
    d. 1 mm
    e. none of the above

23. Which of the following are types of light microscopes? (1) Ultraviolet; (2) fluorescence; (3) electron; (4) darkfield; (5) phase-contrast
    *a. 1, 2, 4, and 5 only
    b. 2, 3, and 4 only
    c. 3, 4, and 5 only
    d. 1 and 2 only
    e. All of the above (1 through 5)

24. Which of the following men proved by experimentation the germ theory of disease?
    a. Leeuwenhoek
    b. Needham
    c. Spallanzani
    *d. Koch

25. The man given credit for the first description of bacteria is
    a. Pasteur
    *b. Leeuwenhoek
    c. Gram
    d. Koch

26. On a compound microscope the eyepiece on top is also called the
    a. upper lens
    *b. ocular
    c. objective
    d. resolving power
    e. paraboloid condenser

27. The 100✕ objective lens with a 10✕ ocular lens provides an overall magnification of
    a. 440
    *b. 1,000
    c. 10,000
    d. 40
    e. 10

28. Which microscope is useful for virology?
    a. Light
    b. Ultraviolet
    c. Fluorescence
    *d. Electron
    e. Darkfield

29. The book of Leviticus was probably the first book on
    a. the existence of pathogens

b. methods of ancient medicine
*c. public health
d. prayers on health
e. none of these

30. Who is credited with developing the modern techniques of nursing?
    a. Lister
    b. Hesse
    c. Petri's wife
  *d. Nightingale
    e. Koch

31. What plants are able to return nitrogen from the air to the soil in the form of nitrates that are used by other plants?
    a. Apples
  *b. Peas, peanuts, alfalfa, and clover
    c. Bean sprouts and peanuts
    d. Peat moss, alfalfa, and clover
    e. Green beans, potatoes, and carrots

32. Bacteria capable of converting alcohol to acetic acid are called
    a. animalcules
  *b. *Acetobacter*
    c. agar
    d. saprophytes
    e. legumes

33. The oil used with the oil immersion lens
    a. helps locate the organism
    b. helps with the observation of protozoa, algae, and other large microorganisms
  *c. reduces the scattering of light rays for clearer observation
    d. prevents microorganisms from drying up on the slide

34. Some diseases are caused by poor diet, genetic problems, or infectious agents. Which of the following is the result of a pathogen?
    a. Pneumonia
    b. Rickets
    c. Scurvy
    d. Diabetes
    e. Sickle cell anemia

35. Agar is
    a. an extract from a marine seaweed
    b. a disease-causing microbe
    c. used to solidify the growth medium
    d. the first dish used for microbial growth
  *e. *a* and *c* only

Match the following:
_____ 1. Leeuwenhoek
_____ 2. Koch
_____ 3. Pasteur
_____ 4. Lister
_____ 5. Jenner
_____ 6. Janssen
_____ 7. Tyndall

A. Introduced antiseptic surgery
B. Developed the cowpox vaccine for smallpox
C. Proved a specific microbe causes a specific disease
D. Introduced tyndallization to kill spores
E. Developed the biological theory of fermentation and a vaccine for rabies
F. Developed the compound microscope
G. First described bacteria

Chapter 2
Types of Microorganisms

Objectives

1. State the cell theory
2. Give a function for each part of the eucaryotic animal cell
3. Name a function for each part of the bacterial cell
4. Explain the differences among plant, animal, and bacterial cells.
5. List the characteristics used to classify bacteria
6. State the differences among rickettsias, chlamydias, and mycoplasmas
7. Name several important bacterial diseases
8. List the classes of protozoa and characteristics for classifying them
9. List five pathogenic protozoa
10. State some important characteristics of fungi
11. List five diseases caused by fungi
12. Discuss the important characteristics of procaryotic and eucaryotic algae
13. Discuss the characteristics that make algae different from protozoa and fungi
14. Describe the characteristics used to classify viruses
15. Compare some of the differences between viruses and bacteria
16. List several important viral diseases

Summary and Discussion

Chapter 2 presents the basis of classification of microorganisms. It is important for the student to understand the differences between the procaryotic and eucaryotic organisms because these differences are often the basis for the selectivity of antibiotic activity.

The various types of organisms that make up the kingdom Protista are surveyed. Selected characteristics of the bacteria, viruses, protozoa, fungi, and algae are presented, as well as their ecological contributions, uses, and some of the diseases they cause.

Outline

I. Cells: eucaryotes and procaryotes
  A. Cell theory: Hooke, Scheiden, Schwann, and Virchow
  B. The cell: the fundamental living unit of life
    1. Obtains food and metabolizes it
    2. Reproduces itself
    3. Responds to the environment
  C. Eucaryotic cells: animal, plant, fungi, algae, and protozoa
    1. Cell membrane
    2. Nucleus
      a. Nuclear membrane
      b. Chromosomes
      c. Nucleolus
      d. Nucleoplasm
    3. Cytoplasm
      a. Endoplasmic reticulum, ribosomes
      b. Golgi apparatus
      c. Lysosomes
      d. Mitochondria
      e. Centrioles
      f. Chloroplasts in plants
    4. Cell wall
  D. Procaryotic cells: bacteria and cyanobacteria
    1. Nucleoid
    2. Cytoplasm
    3. Cytoplasmic particles
    4. Cell membrane
      a. Mesosomes
    5. Bacterial cell wall
    6. Capsules
    7. Flagella
    8. Fimbriae, pili
    9. Endospores
  E. Differences between procaryotes and eucaryotes
II. Microbial classification
III. Bacteria
  A. Characteristics
    1. Morphology (grouping)
    2. Staining characteristics
    3. Motility
    4. Colony characteristics
    5. Nutritional requirements
    6. Biochemical activities
    7. Pathogenicity
  B. Rudimentary forms of bacteria
    1. Rickettsias
    2. Chlamydias
    3. Mycoplasmas
IV. Protozoa
  A. Classified by means of locomotion
    1. Cilia
    2. Flagella
    3. Pseudopodia
  B. Diseases caused by protozoa
V. Fungi (mushrooms, molds, yeasts)
  A. Characteristics
  B. Classification of true fungi
    1. Mode of reproduction (sexual, asexual)
    2. Color and arrangement of spores
    3. Shape of spore-bearing structure
    4. Types of hyphae
  C. Commercial importance
  D. Dimorphism
  E. Fungal diseases
    1. Superficial mycoses
    2. Deep-seated and systemic mycoses
VI. Algae and cyanobacteria
  A. Characteristics
    1. Photosynthetic
    2. Multicellular or unicellular
    3. One pathogenic species
  B. Commercial importance
VII. Acellular Infectious Agents
  A. Viruses
    1. Means of classification
      a. Genetic material
      b. Size
      c. Shape
      d. Presence of envelope
      e. Host organism
      f. Type of disease
    2. Bacteriophages
    3. Viruses and genetic changes
    4. Diseases caused by viruses
  B. Viroids and Prions

Suggested Laboratory Exercises

1. Observe a wet mount of a drop of pond water and identify various protozoa and algae.
2. Under a dissecting microscope, examine the molds on bread or other foods and notice the differences in color, sporulating structures, and spores.
3. Examine a wet mount of yeast suspended in water and observe the budding.
4. Make a smear and add Gram's stain to the bacteria in a drop of saliva. Describe the morphology of the bacteria.

Suggested Audiovisual Aids

1. *Biological Laboratory Techniques: Staining.* 16-mm, color, sound, 10 min. (MG)
2. *Biological Laboratory Techniques: Microscope.* 16-mm, color, sound, 11 min. (MG)

3. *Microscopic Life in Soil.* 16 mm, color, sound, 14 min. (STAN)
4. *Microbiology. No. 3, Complex Microorganisms* (AIBS, Part II). 16 mm, color, sound, 30 min. (IU)
5. *Algae.* 16 mm, color, sound, 17 min. (IU)
6. *Bacteria—Invisible Friends and Foes.* Color videotape W–772–VS (HC)
7. *Mold.* Color videotape, 7 min. # 1283F (FFH)
8. *Protozoa.* Color videotape, 30 min. (SSS)

Sample Examination Questions

1. Which of the following are procaryotic?
   a. Molds and yeasts
   b. Protozoa and green algae
   *c. Bacteria, rickettsias, and cyanobacteria
   d. Higher plants
   e. All of the above
2. Which of the following are eucaryotic?
   a. Viruses
   *b. Fungi
   c. Cyanobacteria
   d. Bacteria
   e. All of the above
3. Characteristic of eucaryotic cells is (are)
   a. a nucleus bound by a nuclear membrane
   b. an endoplasmic reticulum
   c. mitochondria
   d. chloroplasts, if chlorophyll is present
   *e. all of the above
4. The major difference between procaryotic cells and eucaryotic cells is
   *a. eucaryotic cells have a well-defined nuclear membrane whereas procaryotic cells have no true nucleus
   b. procaryotic cells have a well-defined nuclear membrane whereas eucaryotic cells do not
   c. eucaryotic cells are usually found in pathogenic bacteria
   d. there is very little difference between them
5. The energy-producing organelle(s) in the eucaryotic cell is/are
   a. the Golgi bodies
   b. the lysosomes
   c. the chloroplasts
   *d. the mitochondria
   e. the endoplasmic reticulum
6. Which of the following is a characteristic of a procaryotic cell?
   a. Multiple chromosomes
   b. Mitochondria
   c. A system of mitosis

*d. The lack of nuclear membrane
   e. A nuclear membrane
7. What organisms could be acid-fast stained?
   a. Any that have flagella
   b. Organisms that cause tetanus
   c. Organisms that absorb gentian violet
   *d. *Mycobacterium* species
   e. *Salmonella* species
8. Hanging drop preparations are useful for
   a. studying bacterial growth
   b. observing chemical differences in cells
   *c. observing bacterial motility
9. A gram-positive bacterium is
   a. red
   *b. violet or purple
   c. green
   d. colorless
10. An organism that has flagella all the way around it is described as
    *a. peritrichous
    b. lophotrichous
    c. fuzzy
    d. monotrichous
    e. amphitrichous
11. Multiplication by elongation of a cell followed by a division of the enlarged cell into two cells is known as
    a. fusion
    b. pleomorphism
    c. separatism
    *d. binary fission
    e. inclusion
12. A minute, highly durable body formed within the cell and capable of developing into a new vegetative cell is
    *a. an endospore
    b. a hard core
    c. a capsule
    d. a fimbria
    e. a nucleus
13. The chemical composition of most capsules is
    a. polysaccharide
    b. polypeptide
    c. mucopeptide
    d. glycopeptide
    *e. *a and b*
14. Two genera of rod-shaped bacteria, *Bacillus* and *Clostridium*, have the ability to develop a specialized structure for survival called the
    a. ribosome
    *b. endospore
    c. mesosome
    d. chromatophore
    e. nucleus
15. Examples of motile bacteria are found among

a. spirilla
b. bacilli
c. cocci
*d. *a* and *b*
e. all of the above

16. Bacteria occur in three forms or shapes: cocci, bacilli, and spirilla. Which of the following is the best description of each?
    a. Spirilla are curved rods, bacilli are flower shaped, and cocci are round.
    *b. Spirilla are spirals, bacilli are rods, and cocci are round.
    c. Cocci, bacilli, and spiral forms are all pathogens.
    d. Cocci, bacilli, and spirilla are seen in a positive "strep" throat culture.
    e. Cocci, spirilla, and bacilli all look alike.

17. The study of form and structure of bacteria is
    a. zoology
    b. microbiology
    *c. morphology
    d. artology
    e. mycology

18. Bacterial cells that are spherically shaped are
    *a. cocci
    b. round
    c. balli
    d. spirilla
    e. bacilli

19. Round bacteria occurring in chains are called
    a. diplococci
    b. staphylococci
    *c. streptococci
    d. sarcinae
    e. gaffkya

20. Amphitrichous organisms have
    *a. flagella on both ends
    b. flagella on one end
    c. one flagellum
    d. less than five flagella
    e. no flagella

21. What shape are bacilli?
    a. Round
    b. Curved rods
    c. No definite shape
    d. Flat with concave surfaces
    *e. Rods

22. Bacteria move by means of
    a. pseudopodia
    *b. flagella
    c. fimbriae
    d. pili
    e. *a* and *c*

23. Which of the following is used for species identification of bacteria?

a. Motility
b. Morphology
c. Biochemical characteristics
d. Gram-staining reaction
*e. All of the above

24. Bacteria are identified by
    a. *Bergey's Manual of Systemic Bacteriology*
    b. waiting until they produce a disease; then you can be sure by the symptoms
    c. the overall size and appearance of the bacterial colony
    d. the size and shape of the organism, the stain reaction, and, if motile, the type of flagellation
    *e. *a, c,* and *d* only

25. Bacteria appear in which of the following shapes?
    a. Cocci
    b. Bacilli
    c. Spirilla
    d. None of the above
    *e. All of the above

26. Tetanus, gas gangrene, and botulism are caused by organisms of genus
    a. *Bacillus*
    *b. *Clostridium*
    c. *Spirillum*
    d. *Sporocina*
    e. *Desulfotomaculus*

27. Certain bacterial species can form a resting body known as a (an)
    a. inclusion
    b. chromatophore
    c. protoplast
    *d. endospore
    e. forespore

28. Structures that provide for motility in bacteria are
    *a. flagella
    b. cilia
    c. pseudopodia
    d. fimbriae

29. The nuclear material of the bacterial cell has been
    *a. shown to exist as a single molecule of DNA
    b. found in two or more chromosomes
    c. found to have a distinct nuclear membrane
    d. shown to be RNA
    e. shown to be ATP

30. Gram stain differentiates between
    a. viruses and bacteria
    *b. gram-positive and gram-negative bacteria
    c. living organisms and dead organisms
    d. plants and animals

31. Stained bacteria means that the

organisms have been colored with a chemical stain to make them easier to see and study. Which of the following is correct?
  a. Staining can be done only by the hanging-drop method.
  *b. Stained smears of bacteria reveal size, shape, and arrangement and the presence of certain internal structures.
  c. Stained bacteria are basically useless
  d. Staining is useful only when using a magnifying glass.
  e. Staining is dangerous when used with pathogens because it will kill the pathogen.

32. Which is a characteristic of a gram-negative organism?
  a. They are blue.
  *b. They retain a red counterstain.
  c. They are violet or purple.
  d. They retain the purple dye complex after a 95% alcohol wash.
  e. They are easily killed by small amounts of penicillin.

33. The acid-fast stain is used to identify organisms of the genus
  *a. *Mycobacterium*
  b. *Bacillus*
  c. *Clostridium*
  d. *Streptococcus*
  e. *Pseudomonas*

34. Which of the following are not procaryotic "protista"?
  a. Cyanobacteria
  *b. Protozoa
  c. Bacteria
  d. Rickettsia
  e. Chlamydia

35. *Bergey's Manual* is used as a reference for
  a. procedures of antibiotic assay
  b. information of disinfectants
  *c. descriptions of bacterial species
  d. methods of performing biochemical tests

36. Which of the following statements reveals the present concept of nuclei in bacteria?
  a. Definite, particulate, easily seen
  b. Not present
  c. Multinucleated cells
  *d. Nuclear material present, but no nuclear membrane

37. Cocci that divide into two or three planes to form irregular clusters of cells are called
  a. diplococci
  b. streptococci
  c. sarcinae
  *d. staphylococci

  e. gaffkya

38. Which of the following means is used to observe living bacteria?
  a. Staining
  *b. Hanging-drop preparation
  c. Smear
  d. Simple stain
  e. Gram stain

39. Elaborate sterilization procedures are required in hospitals and canneries because of
  a. capsules
  *b. endospores
  c. flagella
  d. enzymes

40. The structure providing bacterial cells with rigidity and outlining their shape is the
  a. cell membrane
  *b. cell wall
  c. hard capsule
  d. cytoplasmic membrane

41. The term *staphylococcus* implies the arrangement of
  a. cocci in chains
  b. bacilli in clusters
  c. cocci in cubes
  *d. cocci in clusters
  e. cocci in pairs

42. Which of the following statements is true concerning flagella?
  a. Flagella resemble yeast spores.
  b. Flagella resemble bacterial spores.
  *c. Flagella are useful for bacterial locomotion.
  d. Flagella are gram-positive structures of most yeast cells.

43. Bacteria commonly multiply by
  a. spore germination
  b. conjugation
  c. ascospore formation
  d. budding
  *e. binary fission

44. Rickettsias are usually described as obligate parasites because
  *a. they grow and multiply only within living cells
  b. they can be cultivated on blood agar
  c. they cause disease in humans

45. Why is marked pleomorphism characteristic of *Mycoplasma* (PPLO)?
  a. They possess a flexible wall external to their membrane.
  *b. They lack a rigid cell wall.
  c. Their shape changes as the cells age.

46. Rickettsias cause diseases in humans, to whom they are transmitted by
  a. contact with rodents
  *b. bites of an arthropod vector

c. contact with an infected human

d. eating infected food

e. drinking soda water

47. The smallest microorganisms known are
    a. rickettsias
    b. protozoa
    *c. viruses
    d. bacilli
    e. inclusion bodies

48. Rickettsias are
    a. smaller than most bacteria
    b. obligate intracellular parasites
    c. rod shaped
    d. closely related to gram-negative bacteria
    *e. all of the above

49. Mycoplasmas
    a. are the smallest living organisms capable of growth and reproduction outside of a living host
    b. are also known as PPLO
    c. do not possess a rigid cell wall
    *d. all of the above
    e. none of the above

50. Of what significance are L-forms in fighting disease?
    *a. They can change form to resist unfavorable environments, then change back to their original state afterward.
    b. They are always pathogens.
    c. They are the cause of gastroenteritis.
    d. They are a breed of virus.
    e. They look like the letter L.

51. Rickettsias are generally
    a. larger than bacteria
    b. smaller than viruses
    *c. larger than viruses
    d. the same size as protozoa

52. One of the distinguishing features of viruses is that they are
    *a. intracellular parasites
    b. gram-positive
    c. saprophytic
    d. nonpathogenic

53. The organisms that cause typhus fever are
    a. viruses
    *b. rickettsias
    c. bacteria
    d. molds

54. To which category of microbes do bacteriophages belong?
    a. Bacteria
    *b. Viruses
    c. Rickettsias
    d. Yeasts

55. A virus that infects a bacterial cell is called
    *a. a bacteriophage
    b. an inclusion

c. a chromatid

56. When a bacterial culture is infected with a temperature phage, the culture is said to be
    a. temperate
    *b. lysogenic
    c. stationary

57. To produce exotoxin, a culture of *Corynebacterium diphtheriae* must be
    a. temperate
    *b. lysogenic
    c. stationary

58. Negri bodies are useful in diagnosing
    *a. rabies
    b. measles
    c. smallpox
    d. vaccinia

59. The most common method of reproduction in molds is
    *a. spore formation
    b. binary fission
    c. budding

60. A resistant form developed by many protozoa is called
    a. an endospore
    *b. a cyst
    c. a chlamydospore

61. Why is the brain examined in a dead animal suspected of having rabies?
    *a. For the presence of Negri bodies
    b. For infected brain cells, which turn a light yellow
    c. Sometimes the virus can be seen
    d. For the presence of L-forms
    e. This is of little value

62. Viruses cannot live outside a host because
    a. they are too weak.
    b. they have no ATP-generating system
    c. they have no way to synthesize their own proteins
    d. *a* and *b*
    *e. *b* and *c*

63. Mutated viruses are useful in many ways. The most valuable use is
    *a. as vaccine for pathogenic strains
    b. for human cancers
    c. in the classification of viruses
    d. for control of infection

64. How do viruses reproduce?
    a. Binary fission
    *b. By taking over the cell
    c. Mitosis
    d. Sexually
    e. None of the above

65. The basic technique used to determine virus size is
    a. filtration through graded membranes in which the pore size of the membrane is known

b. high-speed centrifugation in which the size of the virus can be calculated by determining the rate at which the virus particles settle to the bottom of the centrifuge tube
c. direct observation with an electron microscope
*d. all of the above

66. Classification of animal viruses is based on
    a. the type of nucleic acid
    b. their sensitivity to ether
    c. the presence of a limiting membrane
    d. their symmetry (cubical or helical)
    e. the number of capsomeres making up the capsid
    *f. all of the above

67. Fever blisters, chickenpox, and venereal herpes are caused by
    a. kissing
    b. papovaviruses
    c. adenoviruses
    *d. herpes viruses
    e. poxviruses

68. Viruses are obligate intracellular parasites that contain
    a. ATP-generating system
    b. ribosomes for protein synthesis
    *c. RNA or DNA but never both
    d. all components necessary for replication
    e. regeneration system

69. Lysogeny is a condition
    *a. whereby bacteriophage DNA can exist inside a host bacterium without producing mature bacteriophages
    b. whereby a bacteriophage lysogenizes a bacterium
    c. that causes lysis and death of bacteria
    d. whereby mature phage particles are produced
    e. whereby bacteriophage nucleic acid replicates repeatedly within the infected cell

70. Interferon is
    a. a virus that produces a disease
    *b. an antiviral substance produced by cells infected with a virus
    c. a phage that produces lysogeny
    d. cloudy
    e. an agent that carries a disease from one host to another

71. Protozoa are a group of microorganisms that
    *a. are the most simple and primitive animals
    b. are made up of many clusters of cells
    c. are large and can be seen with the naked eye

d. are also called lichens and grow on rocks in various colors
e. grow only in the human intestine

72. True fungi
    *a. lack chlorophyll
    b. contain chlorophyll
    c. use sunlight for growth and development
    d. live within living cells
    e. are autotrophic

73. Algae, the most simple plants,
    a. are found everywhere
    *b. use chlorophyll and therefore require sunlight
    c. lack chlorophyll
    d. are always unicellular

74. The usual method of reproduction of yeasts is
    a. fertilization
    b. spores
    *c. budding
    d. binary fission
    e. mitosis

75. Dermatophytes
    a. produce infections limited to superficial layers of the body
    b. may cause ringworm of the scalp
    c. can clear spontaneously
    d. can remain as a low-grade, chronic infection
    *e. all of the above

76. *Candida albicans* causes
    a. ringworm
    b. asthma
    c. histoplasmosis
    d. hair infections
    *e. thrush

77. The antibiotic penicillin is obtained from
    a. yeasts
    *b. molds
    c. protozoa
    d. slime molds
    e. filamentous bacteria

78. Molds and yeasts
    a. are procaryotic
    *b. are eucaryotic
    c. can be either procaryotic or eucaryotic

79. The most important function of free-living protozoa is
    *a. maintaining the balance of nature
    b. the control of viruses
    c. producing exotoxin
    d. restoring free N_2 to air
    e. speeding up reactions

80. Protozoa are classified by their means of
    a. reproduction
    *b. locomotion
    c. respiration
    d. elimination

81. The following may occur after phage infection
   a. virulent phage will mature with the lytic cycle
   b. temperate phage produces lysogeny
   c. a lysogenic culture may be induced to maturation by ultraviolet light
  *d. all of the above
   e. none of the above

Chapter 3
Introduction to Chemistry of Life

Objectives

1. Differentiate among elements, atoms, molecules, and compounds
2. Describe an acid-base reaction
3. Discuss the importance of water in biochemical reactions
4. List the characteristics of monosaccharides, disaccharides, and polysaccharides
5. Describe four main types of biochemical molecules
6. Distinguish between organic and inorganic compounds
7. Discuss the structure of carbohydrates, fats, proteins, and nucleic acids and their breakdown products
8. Describe the role of enzymes in metabolism
9. Discuss how DNA directs cellular activities

Summary and Discussion

This chapter introduces the student to the basic concepts and vocabulary of chemistry. Elements, atoms, molecules, and compounds as well as bonding types (ionic, covalent, polar) are illustrated. The importance of water, hydrolysis, and acid-base reactions are discussed. The structure, characteristics, and metabolites of carbohydrates, lipids, proteins, and nucleic acids are emphasized so that the function of enzymes and metabolic reactions may be better understood.

Outline

I. Basic chemistry
  A. Atoms, molecules, elements, compounds
    1. Atomic weight, atomic number
    2. Isotopes
  B. Chemical bonding
    1. Ionic bonds, cations, anions
    2. Covalent bonds

  C. Importance of water
    1. Polar bonds
    2. Solvent properties
    3. Polarity
    4. Buffer
    5. Hydrolysis, dehydration synthesis
  D. Solutions, acids, bases, salts, pH
II. Basic organic chemistry
  A. Carbon bonds
  B. Ring, cyclic compounds
III. Biochemistry
  A. Carbohydrates
    1. Monosaccharides
    2. Disaccharides
    3. Polysaccharides
  B. Lipids
    1. Simple lipids: fats, oils, waxes
    2. Compound lipids: phospholipids, glycolipids
    3. Derived lipids: steroids, cholesterol, vitamins
  C. Proteins
    1. Protein structure
    2. Primary, secondary, tertiary, quaternary protein structure
    3. Enzymes
  D. Nucleic acids, DNA, RNA
    1. Function and structure
    2. DNA replication
    3. Protein synthesis

Suggested Laboratory Exercises

Show the effects of temperature and pH changes on enzymes during the growth of bacteria such as *Escherichia coli* and *Serratia marcescens*. Show how the enzymes in saliva breakdown starches to disaccharides and monosaccharides.

Suggested Audiovisual Aids

1. *Protein Structure and Function*. Media Guild, 11526 Sorreto Valley Rd., San Diego, CA 92121. 16-mm, color and sound.
2. *Evolution by DNA* (1976). Document Associates, Inc., 58 Fenwood Rd., Boston MA 02115, 16-mm color and sound.
3. *The Chemistry of Life*. Color videotape EP–2066V (EI)
4. *Functional Chemistry in Living Cells*. Color videotape SC–770243 (PLP)

Sample Examination Questions

1. Macromolecules constructed by covalently bonding sugars are called

a. proteins
*b. polysaccharides
c. lipids
d. nucleic acids
e. none of the above

2. The chloride ion carries a net charge of −1 because it has more
   a. protons than neutrons
   *b. electrons than protons
   c. neutrons than electrons
   d. protons than electrons
   e. neutrons than protons

3. The glucose molecule's atoms are bonded by which type of chemical bond?
   a. Ionic
   *b. Covalent
   c. Polar ionic
   d. Hydrogen
   e. Nonpolar ionic

4. The atoms in the sodium chloride molecule are bonded by which type of chemical bond?
   *a. Ionic
   b. Covalent
   c. Hydrogen
   d. Glycosidic
   e. Peptide

5. Ionic bonds differ from covalent bonds in that ionic bonds
   a. involve the sharing of a pair of electrons
   b. involve weak attractive forces between two partially charged atoms
   c. involve weak attractive forces between two molecules of different masses
   *d. involve attractive forces between two ions of opposite charge
   e. none of the above statements are true

6. Isotopes of an atom have
   a. the same number of protons but different number of electrons
   b. the same number of protons and neutrons
   *c. the same number of protons but different number of neutrons
   d. the same number of neutrons and electrons
   e. the same number of protons and electron orbitals

7. The branch of chemistry that studies covalently bonded carbon-containing compounds is called
   *a. organic chemistry
   b. inorganic chemistry
   c. biochemistry
   d. physical chemistry

8. Molecules that produce hydroxyl ions when dissolved in water are called

a. buffers
*b. bases
c. acids
d. salts

9. Lipids are commonly found, along with proteins, as part of
   *a. plasma membranes
   b. cell walls
   c. ribosomes
   d. nuclei
   e. solutions

10. Lipid molecules are composed of
    a. amino acids
    *b. glycerol and fatty acids
    c. sugars
    d. nucleotides
    e. only carbon atoms

11. Carbohydrate molecules include
    a. sugar
    b. starch
    c. cellulose
    d. lactose
    *e. all of the above

12. The order, or sequence, of amino acids in a protein molecule constitutes its
    *a. primary structure
    b. secondary structure
    c. tertiary structure
    d. active site
    e. specificity

13. Amino acids are the subunits of which compound?
    a. Carbohydrates
    b. Fats
    c. Phospholipids
    *d. Proteins
    e. Nucleic acids

14. Proteins that speed up the rate of biological reaction are called
    a. proteases
    b. transferases
    *c. enzymes
    d. mucopolysaccharides
    e. phospholipids

15. Digestion is equivalent to
    *a. hydrolysis
    b. dehydration synthesis
    c. metabolism
    d. nutrition
    e. polymerization

16. The function of enzymes is to
    a. catalyze a reaction faster than the spontaneous rate
    b. lower the energy required for the reaction to occur
    c. cause a specific change on a substrate
    *d. all of the above

17. DNA differs from RNA in that DNA

contains which nitrogen base and sugar combination not found in RNA?
- a. Adenine and ribose
- b. Thymine and ribose
- c. Uracil and deoxyribose
- *d. Thymine and deoxyribose
- e. Guanine and ribose

18. All of the following characterize DNA except
- *a. it is replicated on the ribosome
- b. its shape is described as a double helix
- c. it contains the cell's protein code
- d. it may mutate
- e. it can produce exact copies of itself

19. Translation is best described as the direction of
- a. mRNA synthesis by DNA
- b. tRNA synthesis by DNA
- *c. protein synthesis by mRNA
- d. DNA synthesis by mRNA
- e. mRNA synthesis by tRNA

20. RNA is a polymer of
- a. triglycerides
- b. carbohydrates
- c. peptides
- *d. nucleotides
- e. amino acids

21. Polysaccharides are polymers of
- a. nucleotides
- b. amino acids
- *c. glucose
- d. fatty acids
- e. sucrose

Chapter 4
Physiology of Microorganisms

Objectives

1. List the nutritional types of bacteria
2. Discuss how the nutritional types fit into the biosphere
3. State the meaning of autotroph, heterotroph, chemotroph, and phototroph
4. Define producers, consumers, and decomposers
5. Describe and give an example of catabolism, anabolism, respiration, and photosynthesis
6. List six uses for energy in the cell
7. Draw and label the bacterial growth curve
8. List the reasons bacteria die during the death phase
9. Describe the bacterial chromosome
10. List and describe five ways by which the genetic constitution of bacteria can be changed

Summary and Discussion

In Chapter 4 the students are presented with the complex concepts of physiology, nutrition, metabolism, and genetics of all living cells. These concepts can best be presented by comparing the physiological and ecological activities of all organisms, noting that some types of bacteria fit into each of the nutritional and metabolic groups. The section on bacterial genetics can easily be related to the genetics of any cell. It should be noted that these changes—mutation, transformation, transduction, conjugation, and lysogenic conversion—probably occur every day in the human intestine. The transfer of genes by transformation, using *Escherichia coli* genes will make a good topic for the discussion of genetic engineering in humans. The population growth curve also can be an interesting discussion topic by relating it to animal and human populations.

Outline

I. Nutrition
  A. Nutritional requirements
    1. Six major elements: carbon, hydrogen, oxygen, nitrogen, phosphorus, and sulfur
    2. Other elements necessary in lesser amounts
    3. These elements are used to build the carbohydrates, fats, proteins, and nucleic acids of living cells
  B. Nutritional types
    1. Heterotrophic organisms
      a. Chemoorganotrophs
      b. Photoorganotrophs
    2. Autotrophic organisms
      a. Chemolithotrophs
      b. Photolithotrophs
        (1) Photosynthesis
II. Metabolic enzymes
  A. Factors affecting enzyme activity
    1. Substrate, enzyme specificity
    2. Optimum conditions
    3. Cofactors, coenzymes
  B. Naming of enzymes
    1. Endoenzymes, exoenzymes
    2. Substrate types
      a. Proteases
      b. Lipases
      c. Carbohydrates
    3. Reaction types
      a. Hydrolases
      b. Polymerases
  C. Inhibition of enzymes
    1. Change optimum conditions

2. Mineral ions
3. Similar substrates
III. Cellular metabolism
  A. Catabolism and anabolism
  B. Energy production (catabolism)
    1. Cellular respiration of glucose
      a. Glycolysis
      b. Citric acid cycle
      c. Electron transport system
      d. Alternate pathway
    2. Fermentation
    3. Aerobic oxidation by chemolithotrophs
    4. Anaerobic respiration by chemotrophs
  C. Metabolic biosynthesis
    1. Energy conversion
      a. Photosynthesis
      b. Chemosynthesis
IV. Microbial growth
  A. Factors that influence growth
    1. Temperature
    2. pH
    3. Moisture
    4. Nutrients
    5. Other organisms
  B. Culture media
  C. Technique for determining the number of bacteria per milliliter
    1. Counting chamber
    2. Electronic cell counter
    3. Turbidity
    4. Viable plate count (used to count the number of living bacteria per milliliter)
  D. Population growth curve
    1. Lag phase
    2. Logarithmic growth phase
    3. Stationary phase
    4. Death phase
      a. Factors influencing death phase
V. Bacterial genetics
  A. Chromosomes, genes, and DNA
  B. Changes in bacterial genetic construction
    1. Mutations
    2. Lysogenic conversion
    3. Transduction
    4. Transformation
    5. Conjugation
  C. Genetic engineering

Suggested Laboratory Exercise

Perform the plate count technique as shown in Figure 4–7 in the text. Dilute a culture of *Escherichia coli* three times by transferring 1 ml of solution to 99 ml of water. Then culture 0.1 ml from each dilution bottle on a nutrient agar plate. Spread the organisms with a sterile glass rod for complete coverage. Incubate at 37°C for 24 hours. Count the number of colonies on each plate that has between 30 and 300 colonies. Calculate the number of organisms per milliliter of original culture.

Suggested Audiovisual Aids

1. *Microorganisms: Beneficial Activities.* 16–mm, color, sound, 15 min. (IU)
2. *Metabolism: Structure and Regulation.* Color videotape EV2152V (EI)
3. *Energy Cycles in the Cell.* 16 mm, color, sound, 17 min. (MG)
4. *Gene Action.* 16–mm, color, sound, 16 min. (EB)
5. *Gene Engineers.* NOVA videotape, 57 min. (TLV)

Sample Examination Questions

1. All of the following are essential nutrients for a living cell *except*
  a. water
  b. carbon source
  c. certain inorganic ions
  d. a nitrogen source
  *e. rubber
2. Autotrophs are organisms that
  a. use organic compounds as nutrients
  b. use light as a source of energy
  *c. use inorganic carbon as a source of energy
  d. use any chemical as a source of energy
  e. need hydrogen compounds
3. Bacteria that can transform inorganic matter into proteins, fats, and carbohydrates are called
  a. saprophytes
  b. heterotrophs
  c. lithotrophs
  *d. autotrophs
  e. parasites
4. Organisms that use inorganic compounds for nutrients and synthesis are termed
  a. heterotrophs
  *b. autotrophs
  c. phototrophs
  d. saprophytes
5. Organisms that live on dead, organic matter are called
  *a. saprophytes
  b. parasites
  c. phototrophs
  d. aerobes

e. autotrophs

6. Photosynthetic organisms
   a. use chemicals for energy
   *b. use sunlight for energy
   c. use other cells for energy
   d. none of the above
   e. all of the above

7. Heterotrophic bacteria obtain their energy from the oxidation of
   *a. organic molecules
   b. inorganic molecules
   c. phosphates

8. Chemolithotrophs are organisms that
   *a. obtain nutrients from inorganic substances
   b. rely on ready-made organic food
   c. are saprophytic
   d. are parasitic
   e. contain chlorophyll

9. What is the function of a saprophyte in nature?
   a. Changes sunlight into ATP
   b. Lives off a living organism
   c. Changes inorganic compounds into organic compounds
   d. Produces vitamin K in the intestine
   *e. Decays organic material

10. Organisms that use light as an energy source are called
    a. leukocytes
    *b. phototrophs
    c. parasites
    d. saprophytes
    e. organotrophs

11. The saprophytic fungi and bacteria that live on dead plant and animal material are
    *a. heterotrophs
    b. autotrophs
    c. phototrophs

12. Catabolism could best be described as
    a. buildup metabolism
    *b. breakdown metabolism
    c. overall metabolism
    d. feline murder

13. Basic nutrients required by cells include
    a. carbon and nitrogen sources
    b. certain inorganic ions
    c. essential metabolites for some
    *d. all of the above
    e. none of the above

14. The action of an enzyme
    a. is to buffer pH
    b. is to provide nutrition
    *c. is to speed reactions without changing them
    d. has not been determined

15. In which phase of the growth curve is the greatest population density reached?
    a. Logarithmic
    b. Lag
    *c. Stationary
    d. Death

16. Bacterial populations are usually expressed as the logarithm of the number of living bacteria per
    a. loopful
    *b. ml (milliliter)
    c. tube
    d. slide

17. Which of the following techniques would most likely be employed to obtain data for constructing a typical bacterial growth curve?
    *a. Colony count
    b. Autoanalyzer
    c. Cell nitrogen
    d. Turbidimetric

18. Which combination of data is required to plot the conventional growth curve of bacteria?
    *a. Log of number of living bacteria versus time
    b. Log of number of living and dead bacteria versus time
    c. Log of number of generations versus time
    d. Log of generation time versus time

19. The portion of a bacterial growth curve in which bacteria are *not* reproducing but are very active metabolically is called the
    *a. lag phase
    b. stationary phase
    c. logarithmic phase
    d. steady phase

20. The period of most rapid growth of bacteria where the growth *rate* is constant is termed the
    a. lag phase
    *b. logarithmic phase
    c. maximum death phase
    d. death phase

21. The common method of reproduction in bacteria is
    a. spore formation
    b. mitosis
    *c. binary fission

22. The stationary phase occurs when
    a. bacteria are inoculated into medium and reproduction has not yet begun
    b. the rate of reproduction is rapid
    *c. the reproduction and death rates are equal
    d. reproduction has slowed down
    e. reproduction has stopped

23. Mutations occur at what level?
    a. The RNA level
    *b. The DNA level

c. Both a and b
24. DNA
    *a. controls cellular activity
    b. is the same as RNA
    c. acts as the "powerhouse" of the cell
25. Mutations
    a. help the microorganism adjust to a new environment
    b. occur naturally
    c. can be induced
    *d. all of the above
26. Transfer of genetic material from a donor to a recipient bacterial cell with bacteriophage as a carrier is
    a. conjugation
    b. transcription
    *c. transduction
    d. transformation
    e. allegation
27. A phage that causes lysis in infected bacteria is called a
    *a. virulent phage
    b. prophage
    c. temperate phage
    d. bacterial phage
    e. leptogenic bacterium
28. Which is *not* a way of transferring genetic material?
    a. Transformation
    b. Transduction
    *c. Translation
    d. Conjugation
    e. Phage conversion
29. Genetic material in bacteria can be passed from one cell to another by
    a. transformation
    b. transduction
    c. conjugation
    *d. all of the above
30. Genetic material in transformation was found to be
    *a. DNA
    b. RNA
    c. protein
31. In transduction, DNA is carried to a recipient bacterium by
    a. simple diffusion
    *b. bacterial virus
    c. cellular contact
    d. conjugation
32. Transduction and transformation differ in that transduction
    a. involves DNA
    b. is restricted to the pneumococci
    *c. requires a phage
    d. is a sexual process
33. Which of the following denotes an accidental change in the DNA structure and the characterics of bacteria?

*a. Mutation
    b. Transduction
    c. Transformation
    d. All of the above
34. Which type of genetic change requires a pilus bridge?
    a. Transformation
    b. Mutation
    c. Transduction
    *d. Conjugation
    e. Phage conversion
35. Mutations occur in bacteria as a result of mistakes in the replication of DNA. They occur
    *a. randomly and spontaneously
    b. orderly and predictably
    c. only once or twice
    d. never
    e. very seldom
36. DNA
    a. exists in the cell as a double helix
    b. uses its message for the synthesis of specific proteins
    c. carries messages that control the activities of the cell
    *d. all of the above
    e. none of the above
37. Viruses that can remain dormant in host cells are described as
    a. virulent
    b. domiciles
    *c. temperate
    d. resistive
    e. fast acting
38. The breakdown of complex large molecules into small molecules is called
    a. anabolism
    b. digestion
    c. catabolism
    d. *a* and *b*
    *e. *b* and *c*
39. Which is the process in which oxygen participates in an energy-yielding reaction?
    a. Anabolism
    b. Catabolism
    c. Anaerobic respiration
    *d. Aerobic respiration
    e. Fermentation
40. Bioluminescence is
    a. the emission of light by living organisms in the ocean
    b. a process by which heat is lost
    *c. both *a* and *b*
    d. neither *a* nor *b*
41. The most commonly used method of measuring cell mass is
    a. the membrane filter technique
    b. a colony count

*c. turbidity
  d. a viable cell count
42. Beta-hemolytic streptococci produce a
    _____ zone around colonies on a blood
    agar plate.
    *a. clear
    b. green
    c. red
    d. blue
    e. yellow
43. As a rule, organisms do not invade the
    bladder because it is frequently
    a. sterile
    b. protected by other organisms
    *c. flushed by acidic urine
    d. *a* and *b*
    e. none of the above

Chapter 5
Control of Microbial Growth

Objectives

1. List three reasons why microbial growth
   must be controlled
2. Define sterilization, disinfection,
   bactericidal agents, and bacteriostatic
   agents
3. Differentiate between sterilization,
   pasteurization, and lyophilization
4. Describe aseptic, antiseptic, and sterile
   techniques
5. List the factors that influence the growth
   of microbial life
6. Describe the following types of
   microorganisms: psychrophilic,
   mesophilic, thermophilic, halophilic,
   haloduric, alkaliphilic, acidophilic, and
   barophilic
7. List several factors that influence the
   effectiveness of antimicrobial methods
8. List the common *physical* antimicrobial
   methods
9. List the common *chemical* antimicrobial
   compounds
10. Describe the mode of action of
    sulfonamide drugs on bacteria
11. Describe the action of penicillin on
    bacteria
12. List four reasons why antibiotics should
    be used with caution

Summary and Discussion

In Chapter 6 the important methods used to
control the growth of microorganisms are discussed. The physical methods include the use
of temperature, pressure, drying, change in os-
motic pressure, radiations, ultrasonic waves,
and filtration. The chemical methods include
the use of disinfectants and antiseptics to inhibit the growth of microorganisms by injuring
cell membranes, inactivating enzymes, or dam-
aging genetic material. The study of chemotherapy should be stressed, as well as how
chemotherapeutic agents destroy pathogens
but not the host cells.

Outline

I. Control of growth of microorganisms
   A. Importance of controlling microbial
      growth
      1. To prevent and control infectious
         disease
      2. For food preservation
      3. In industrial processes
      4. For pure culture research
   B. Definitions
      1. Infectious disease
      2. Contamination
      3. Sterilization
      4. Disinfection
         a. Pasteurization
      5. Microbicidal agents
         a. Bactericidal agents
         b. Fungicides
         c. Virucides
         d. Germicides
      6. Microbistatic agents
         a. Bacteriostatic agents
      7. Asepsis
         a. Aseptic technique
         b. Antiseptic technique
      8. Sterile technique
II. Factors influencing microbial growth
    A. Temperature
       1. Psychrophiles
       2. Mesophiles
       3. Thermophiles
    B. Moisture
       1. Desiccation
       2. Lyophilization
    C. Osmotic pressure
       1. Plasmoptysis
       2. Plasmolysis
       3. Halophilic organisms
       4. Haloduric organisms
    D. pH (acidity)
       1. Acidophiles
       2. Alkalophiles
    E. Barometric pressure
    F. Gases
       1. Aerobes
       2. Anaerobes: obligate and facultative

III. Antimicrobial methods
  A. Physical methods
    1. Heat
      a. Thermal death point
      b. Thermal death time
      c. Dry heat
      d. Moist heat
      e. Pressurized steam
    2. Cold
    3. Desiccation
    4. Radiation
    5. Ultrasonic waves
    6. Filtration
  B. Chemical methods
    1. Antisepsis
    2. Characteristics of a good antimicrobial chemical
    3. How antimicrobial chemicals work
      a. Injure cell membranes
      b. Inactivate enzymes
      c. Damage genetic material
IV. Chemotherapy
  A. History
  B. Characteristics of a good chemotherapeutic agent
  C. How chemotherapeutic agents work
    1. Competitive enzyme inhibitors
    2. Antibiotics
  D. Side effects of chemotherapy

Suggested Laboratory Exercises

1. *Antiseptic effectiveness test.* Using a sterile cotton swab and a nutrient agar plate, cover the surface of the plate with a pure culture of *Staphylococcus aureus.* Streak another plate with *Escherichia coli.* Dip small sterile filter paper disks into various disinfectants, antiseptics, or mouthwashes and apply the disks to the surface of the inoculated agar plates as shown. Label each disk on the agar side of the plate. Incubate 24 hours at 37°C. Measure the zone of inhibition around each disk.

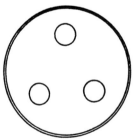

2. *Antibiotic susceptibility test.* On nutrient agar plates that have been inoculated with *Staphylococcus aureus* or *Escherichia coli,* place disks impregnated with various antibiotics, as shown. Incubate the plates at 37°C for 24 hours. Measure the zone of inhibition around each disk. Note any antibiotic resistant colonies near the disks.

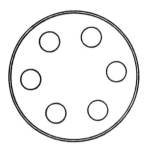

Suggested Audiovisual Aids

1. *Sanitation, Disinfection, and Sterilization for Surgery.* 16-mm, color, sound, 11 min. (NMAVC)
2. *Asepsis.* 16-mm, color, sound, 29 min. (SEF)
3. *Destruction and Inhibition of Microorganisms.* Black and white, 3–part videotape, 90 min. (TS)
4. *Testing of Antimicrobial Agents.* Color videotape, 13 min. (ASM)

Sample Examination Questions

1. Desiccation is a process of
   a. pasteurization
   *b. dehydration
   c. disinfection
   d. sterilization
   e. hydration
2. The word used to describe an agent that will kill bacteria is
   *a. bactericidal
   b. bacteriostatic
   c. fungicidal
   d. allergic
   e. homeostatic
3. The thermophiles grow best in what conditions?
   a. In temperatures near 0°C
   b. All organisms are thermophiles
   *c. In temperatures near 45° to 50°C
   d. At body temperature
   e. They can survive any temperature
4. Organisms that have their optimal growth at temperatures below 15°C are
   a. mesophiles
   *b. psychrophiles
   c. thermophiles
   d. bacteriophiles
   e. protophiles
5. Bacteria that grow best at temperatures between 15°C and 45°C are termed

*a. mesophilic
b. psychrophilic
c. thermophilic
d. aerobic

6. Which of the following methods of food preservation does not establish a microbistatic condition?
   *a. Refrigeration
   b. Deep-freeze storage
   c. Dehydration
   d. Canning

7. Assume that you streaked a nutrient agar plate with your fingers and incubated the plate in your desk drawer. The bacteria most likely to grow would be
   *a. aerobic, heterotrophic, and mesophilic
   b. anaerobic, heterotrophic, and psychrophilic
   c. aerobic, autotrophic, and mesophilic
   d. anaerobic, heterotrophic, and mesophilic

8. Most pathogens grow best at a pH in the vicinity of
   *a. 7
   b. 9
   c. 3
   d. 4.5

9. The process whereby organisms are frozen rapidly then dehydrated is
   a. desiccation
   b. radiation
   *c. lyophilization
   d. pasteurization

10. Complete sterilization by autoclaving requires temperature pressure of
    a. 10 psi @ 100°F
    *b. 15 psi @ 121°C
    c. 100 psi @ 100°C
    d. 25 psi @ 150°C

11. A microorganism that grows either in the absence or the presence of free oxygen is known as
    a. an obligate aerobe
    b. a facultative anaerobe
    c. an anaerobe
    d. a microaerophile
    *e. both *b* and *d*

12. Bacteria that cannot live in the presence of air are known as
    a. aerobes
    b. facultative anaerobes
    *c. obligate anaerobes
    d. microaerophilics
    e. suboxydans

13. What is the purpose of pasteurization?
    a. To prevent the separation of milk and cream
    b. To sterilize the milk
    *c. To kill 90% of the bacteria

d. To kill the spores in milk
e. To filter out pathogens

14. Separation of bacteria from the suspending fluids is accomplished through
    a. sonic vibration
    b. desiccation
    c. pasteurization
    d. suspenderization
    *e. filtration

15. Which of the following is not a method of sterilization?
    a. Moist heat, such as steam
    *b. Using a disinfectant
    c. Boiling water
    d. Dry heat
    e. Radiation

16. The most complete sterilization, which includes killing of endospores, is best accomplished by
    a. boiling
    b. dry heat
    *c. steam under pressure
    d. filtration

17. Factors affecting sterilization are
    *a. time, temperature, and pressure
    b. filtration, time, and cellular constituents
    c. photoreaction, time, and boiling the medium

18. Ultrasound is a technique used for
    a. sending messages
    b. disinfection
    c. radiation
    d. ectopic beats
    *e. sterilization

19. Who was given credit for the first antiseptic surgery?
    a. Pasteur
    b. Semmelweis
    c. Koch
    *d. Lister

20. Which is used to sterilize powders, gauze dressing, and glassware?
    a. Autoclave
    b. Boiling water
    c. Live steam
    *d. Dry heat

21. The best method to sterilize instruments that are contaminated by viruses is
    a. ultraviolet light
    b. pasteurization
    *c. autoclaving
    d. boiling
    e. filtration

22. The aim of the disinfection process is
    a. killing of all forms of life
    *b. destruction of disease agents
    c. immobilization of disease agents
    d. mutation of disease agents

23. The most commonly used method of sterilization is
    a. ultraviolet radiation
    *b. heat
    c. fractional
    d. thermal death point
24. The phenol coefficient test shows
    a. that phenol is the most suitable antiseptic for general use
    b. that no bacteria can resist 5% aqueous phenol for a half hour
    *c. the relationship between other compounds and phenol in comparable bactericidal activity
    d. the pronounced bacteriostatic activity of weak solutions of phenol
    e. the pronounced fungistatic activity of phenol
25. Most skin cancers are caused by what environmental factor?
    a. Smog
    b. Wind
    *c. The sun's ultraviolet light radiation
26. The most effective way to disinfect drinking water at present is
    a. fluoridation
    *b. chlorination
    c. filtration
27. Dressings to be used in surgical procedures should be sterilized by
    a. soaking
    b. Cidex spray
    *c. autoclaving
    d. boiling
    e. hot water bath
28. Which is not a factor to be considered in disinfection?
    a. Time
    *b. Rate of cell division
    c. Concentration
    d. Nature of surrounding area
    e. Temperature
29. What is the most effective concentration of ethyl alcohol?
    *a. 70% solution
    b. 20% solution
    c. 80% solution
    d. 100% solution
    e. 10% solution
30. Lister was the first person to use an antiseptic in surgery; which one did he use?
    a. Alcohol
    b. "Sulfa"
    *c. Carbolic acid
    d. Ascorbic acid
31. Bacteriostasis refers to
    *a. inhibiting the growth of bacteria
    b. killing bacteria

c. inhibiting spores
d. asepsis
e. killing viruses
32. As the temperature increases, the effectiveness of a disinfectant generally
    a. decreases
    *b. increases
    c. stays the same
33. The rate at which a disinfectant works is dependent on
    a. concentration
    b. time
    c. temperature
    *d. all of the above
    e. none of the above
34. Soaps and detergents are
    *a. effective both as cleaning agents and in their action against bacteria
    b. ethylene oxide
    c. totally ineffective in their action against bacteria
35. An antiseptic is a chemical substance that
    a. destroys disease organisms
    b. is not effective against bacteria spores
    c. prevents growth by either inhibiting or destroying microorganisms
    d. is used on nonliving objects
    *e. reduces bacterial numbers to safe levels on living tissues
36. Paul Ehrlich, one of the early investigators in the field of chemotherapy, was noted for his work on
    a. sulfonamides
    *b. arsenic compounds
    c. penicillin
    d. Aureomycin
37. *Staphylococcus aureus* has become resistant to which of the following?
    a. DDT
    b. Sodium pentothal
    *c. Penicillin
38. The destruction of bacteria by the sulfonamide drugs is due to
    a. coagulation of protoplasm
    b. dehydration of protoplasm
    c. dissolving of protoplasm
    *d. interference with nutrition
39. The preparation that led to the discovery of the sulfonamides was a red dye known as
    a. safranin
    b. iodine
    *c. Prontosil
    d. gentian violet
    e. Bismarck brown
40. Penicillin acts by
    *a. inhibiting cell wall synthesis
    b. inhibiting spore formation
    c. injuring the cell membrane

d. inhibiting RNA synthesis
e. inhibiting protein synthesis

41. Waksman is given credit for discovering
   a. penicillin
   b. "sulfa"
   *c. streptomycin
   d. 606
   d. PABA

42. A method of determining whether an organism is sensitive to an antibiotic is
   a. the Gram stain
   b. the Ziehl–Neelsen method
   *c. the disk inhibition technique
   d. the oxidase test
   e. the Neufeld test

43. Chemotherapy is the treatment of disease
   a. with radiation
   *b. with chemicals
   c. with surgery
   d. with isolation
   e. with aseptic

44. The person who discovered penicillin was
   *a. Sir Alexander Fleming
   b. Dr. Selman A. Waksman
   c. William Penn
   d. Dr. De Bakey
   e. Paul Ehrlich

45. Paul Ehrlich is noted for discovering the effectiveness of an arsenic compound called
   a. Ketlin
   b. sulfa drug
   c. iodine
   *d. salvarsan

46. Salvarsan was used in the treatment of
   a. dysentery
   *b. syphilis
   c. gout
   d. gonorrhea

47. One of the ways antibiotics work against pathogens is to
   a. inhibit flagella
   b. enhance conjugation
   *c. inhibit protein synthesis
   d. rearrange the cytoplasm
   e. destroy cilia

48. Penicillin is
   a. bacteriostatic
   b. bacteriophilic
   c. a protozoan
   *d. bactericidal
   e. bacteriogenic

49. Indiscriminate use of antibiotics may cause
   a. allergy to antibiotics
   b. mutant pathogens
   c. resistant pathogens
   d. destruction of normal flora
   *e. all of above

50. *p*H is a measurement of
   a. alkalinity
   b. acidity
   c. temperature
   *d. *a* and *b*
   e. all of the above

51. An alcohol that is too toxic to take internally but is used externally on the skin is
   a. formalin
   b. ethylene oxide
   *c. isopropyl
   d. ethyl

52. Lysol and carbolic acid are
   a. mercury salts
   b. peroxides
   *c. phenolics
   d. formalins
   e. ethylene oxides

53. An oxidizing agent frequently used to cleanse wounds is
   a. a solution of mercury salts
   b. formalin
   *c. hydrogen peroxide
   d. detergent or soap
   e. none of the above

54. Who purified penicillin?
   a. Ehrlich
   b. Fleming
   c. Uster
   *d. Florey and Chain
   e. Pasteur

55. When red blood cells are suspended in a hypotonic solution they tend to swell and burst. This is called
   *a. hemolysis
   b. crenation
   c. plasmolysis
   d. plasmoptysis

Chapter 6
Human and Microbial Interactions

Objectives

1. Discuss the importance of indigenous microflora and where it is found
2. List four types of symbiotic relationships
3. Differentiate between mutualism and commensalism and give examples of each
4. Describe one parasitic relationship
5. Discuss the factors related to the pathogenicity of microbes
6. Describe the ecological interrelationships of plants, animals, and microorganisms

Summary and Discussion

The ecological relationship among organisms is a very important concept to stress for students.

They should know some of the many benefits humans receive from the presence of microorganisms in the soil and on the human body. The factors that influence the growth of these microbes and determine where they will grow are important. One should stress that some of the beneficial indigenous microflora can become opportunists that cause disease and wound and burn infections, if they get into the wrong places.

The important concepts and terms pertaining to free-living, saprophytic, parasitic, and pathogenic organisms should be discussed.

Outline

I. Indigenous microflora of humans
   A. Skin
   B. Mouth
   C. Ear and eye
   D. Respiratory tract
   E. Urogenital area
   F. Gastrointestinal tract
II. Symbiotic relationships
   A. Definition of symbiosis
   B. Mutualism
      1. Synergism
   C. Commensalism
   D. Neutralism
      1. Antagonism
   E. Parasitism
      1. Infestation
      2. Infection
      3. Pathogens
      4. Opportunists
      5. Nonpathogens
III. Microbial ecology
   A. Cycling of nutrients
      1. Nitrogen cycle
   B. Factors affecting types and amounts of microbes

Suggested Laboratory Exercises

1. Using a sterile cotton swab, touch the back of the throat and streak the organisms on a blood agar plate by rolling the swab in one area. Then, using a flamed loop, streak the organisms for isolated colonies as shown:

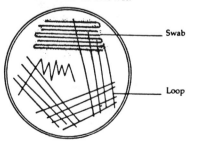

Incubate for 24 hours at 37°C. Observe the differences in the colonies and zones of α- and β-hemolysis. A tiny translucent colony with a large zone of β-hemolysis (clear zone) is probably a β-hemolytic *Streptococcus pyogenes*.

2. Touch five fingers of one hand to the surface of a nutrient agar plate. Wash the hands with soap, then touch the same five fingers to the other half of the agar plate. Incubate the plate at 25°C for 24 hours and observe the number of colonies.

3. Sprinkle a tiny pinch of soil on the surface of a nutrient agar plate. Incubate at 25°C for 24 hours and observe the colonies of microorganisms.

Suggested Audiovisual Aids

1. *Life in a Cubic Foot of Soil.* 16-mm, color, sound, 11 min. (COR)
2. *Isolation and Identification of beta-Hemolytic Streptococci.* 16-mm, color, sound, 16-min. (NMAVC)
3. *The Nitrogen Cycle.* 16-mm, color, sound, 17 min. (UNITED)
4. *Microoganisms—Beneficial Activities.* 16-mm, color, sound (IU)
5. *Parasitism.* 16-mm, color, sound, 17 min. (EB)

Sample Examination Questions

1. Why are the element cycles necessary for human welfare?
   a. To return organic materials to the soil
   b. Because without these cycles the world would have piles of debris
   c. As a mode of travel
   *d. *a* and *b*
   e. None of the above
2. How do microorganisms contribute to the welfare of mankind?
   a. By breaking down organic materials such as leather, garbage, and dead animals and plants
   b. By producing oxygen
   c. By converting inorganic substances into organic substances
   *d. All of the above
   e. None of the above
3. "Fixation of nitrogen" means
   a. production of ammonia from proteins
   b. reduction of proteins to nitrogen gas
   *c. conversion of nitrogen gas to plant nutrients
   d. conversion of nitrates to nitrogen gas
4. Bacteria that "fix" atmospheric nitrogen

in the roots of legumes, in a relationship in which both parties benefit, are
   a. free-living nitrate users
   *b. symbiotic nitrogen fixers
   c. nitrogenase
   d. blue-green algae
5. The microbial population characteristic of some natural environment is called
   a. a pure culture
   b. an abnormal culture
   *c. normal flora
   d. a contaminated culture
6. Which plant is a legume?
   a. Corn
   b. Wheat
   c. Potatoes
   *d. Alfalfa
   e. Cotton
7. Bacterial count is much higher in
   a. sandy soil
   b. clay soil
   *c. well-cultivated soil
8. Which of the following characteristics of various body areas influence the flora of the area?
   a. Pressure
   *b. Temperature, pH, nutrients
   c. Time
   d. Blood flow
   e. Incubation period
9. Indigenous microflora found in the blood are
   a. *Escherichia coli*
   b. *Candida albicans*
   *c. none
10. All are indigenous microflora of the intestine *except*
   a. staphylococci
   b. *Enterobacter*
   c. *Proteus*
   d. *Escherichia coli*
   *e. rabies virus
11. Indigenous microflora that are harmless are called
   a. aerobes
   *b. commensals
   c. opportunists
   d. parasites
   e. cytopathogenic
12. The greatest microbial population on the skin is
   *a. *Staphylococcus*
   b. *Streptococcus*
   c. *Diplococcus*
   d. *Klebsiella*
   e. *Neisseria*
13. A gram-negative facultative bacterium you would except to see in fecal material is
   a. bacteriophage
   b. plasmid

   c. *Klebsiella pneumoniae*
   d. β-hemolytic streptococcus
   *e. *Escherichia coli*
14. In which relationship do the microbe and the host receive benefit from each other?
   a. Commensalism
   b. Opportunism
   c. Pathogenism
   d. Parasitism
   *e. Mutualism
15. An opportunistic organism is
   a. a harmless organism
   b. a parasite
   c. a helpful organism
   *d. a potentially harmful organism
   e. sporadic
16. A microbe that lives on its host and gives no evidence of benefit or harm is known as a (an)
   *a. symbiont
   b. commensal
   c. leech
   d. opportunist
17. The yeastlike fungus *Candida albicans* can cause which of the following diseases?
   a. Rickets
   *b. Thrush, moniliasis
   c. Sickle cell anemia
18. A microorganism that grows either in the absence of or the presence of free oxygen is known as a (an)
   a. obligate aerobe
   *b. facultative anaerobe
   c. anaerobe
   d. parasite
19. An organism that is cultivatable only within living cells is described as
   a. free-living
   *b. an obligate parasite
   c. a saprophyte
   d. a heterotroph

Chapter 7
Microbes versus Humans

Objectives

1. Differentiate among infectious, communicable, and contagious disease
2. List six reasons why an infection may not occur when a pathogen is present
3. Discuss the disease process
4. Define acute and chronic disease
5. State the difference between primary and secondary diseases
6. State the difference between local and generalized infections

7. List three factors associated with the virulence of a pathogen
8. List and discuss eight factors that affect the pathogenicity of bacteria
9. Write the meaning of the following terms: *epidemiology, epidemic, endemic,* and *pandemic*
10. Describe the difference between sporadic and nonendemic diseases
11. List three factors that contribute to an epidemic
12. List six reservoirs of infection
13. List five modes of disease transmission
14. Discuss the procedure for stopping an epidemic

Summary and Discussion

It is very important for the student to gain an understanding of the continual battle waged between the human body and the microorganisms that attempt to invade it. Healthy people are continually protected against the invasion of most pathogens by the skin and mucous membranes and the secretions of these tissues, as well as by the normal flora and their secretions, the phagocytes, the inflammatory response, and the immune response. However, some particularly virulent pathogens can overwhelm these defenses, resist phagocytosis, and invade deep within the body, where they produce toxic materials that cause damage or disease in localized areas or throughout the body.

To clarify many factors and facets of epidemiology, the terms *endemic, nonendemic, sporadic, epidemic,* and *pandemic* must be clearly understood. The modes of transmission of pathogens should be illustrated with many examples of how certain diseases are transmitted. A lively discussion can be developed in the classroom about reservoirs of infection and the prevention and control of epidemics.

Outline

I. Disease and infection
  A. Types of infections
    1. Infectious diseases
    2. Communicable diseases
    3. Contagious diseases
  B. Why infection does not always occur
    1. Wrong place
    2. Antibacterial factors in secretions
    3. Normal flora
    4. Bacteriocidins
    5. Antibodies
    6. Phagocytic activities
  C. The development of infection
    1. Invasion and growth of pathogen
    2. Inflammation
  D. The disease process
  E. Mechanics of disease causation
    1. Virulence of pathogen
      a. Infectivity
      b. Invasiveness
      c. Toxigenicity
    2. Bacterial morphology associated with infectivity
    3. Enzymes associated with invasiveness
      a. Coagulase
      b. Streptokinase
      c. Hyaluronidase
      d. Collagenase
    4. Toxins and enzymes associated with toxigenicity
      a. Hemolysin
      b. Leukocidin
      c. Exotoxins
      d. Endotoxins
  F. Pathogenicity and virulence
II. Epidemiology
  A. Endemic diseases
    1. Tuberculosis
    2. Staphylococcal and streptococcal infections
    3. Virus diseases
  B. Epidemic diseases
    1. Respiratory illnesses
    2. Hospital infections
    3. Venereal diseases
    4. Diseases caused by fecal contamination of water supplies
  C. Pandemics
  D. Sporadic diseases
    1. Tetanus
    2. Gas gangrene
    3. Botulism
  E. Nonendemic diseases
    1. Smallpox
    2. Polio
    3. Diphtheria
  F. Reservoirs of infectious agents
III. Modes of disease transmission
  A. Direct external contact
  B. Venereal or mucus-to-mucus contact
  C. Respiratory or air
  D. Enteric
  E. Blood
    1. Living vectors
      a. Arthropods
    2. Inanimate vectors
      a. Syringes, needles, solutions
      b. Blood-processing equipment
IV. Control of epidemic diseases
  A. Health organizations

B. Reporting cases
C. Education of public
D. Sanitation procedures
E. Elimination of reservoirs
F. Immunization programs
G. Treatment

Suggested Laboratory Exercises

1. *Capsules.* Observe a prepared slide of *Streptococcus pneumoniae* that has been stained to show the presence of capsules.
2. *Hemolysis.* Observe α- and β-hemolysis on the blood-agar plate that has been inoculated with a throat culture and properly streaked for isolated colonies.
3. *Microbes in milk and water.* Streak a dropful of raw milk on a nutrient agar plate using the technique for complete coverage of the plate. Incubate 24 hours at 25°C and count the colonies. Repeat, using a drop of river water.

Suggested Audiovisual Aids

1. *Germfree Animals in Medical Research.* 16-mm, color, sound, 19 min. (NMAVC)
2. *Isolation and Identification of beta-Hemolytic Streptococci.* 16-mm, color with sound 16-min. (NMAVC)
3. *The Triad of Infection.* 16-mm, color, sound, 15 min. (EL)
4. *Communicable Disease.* 16-mm, color, sound, 32 min. (WYA)
5. *The Inflammatory Reaction.* 16-mm, color, sound, 26 min. (LL)
6. *Life in a Cubic Foot of Soil.* 16-mm, color, sound, 11 min. (COR)
7. *For Granted—The Water We Drink.* 16-mm, color, sound, 17 min. (IU)
8. *An Outbreak of Staphylococcus Intoxication.* 16-mm, color, sound, 12 min. (NMAVC)

Sample Examination Questions

1. Factors that are significant in the development of disease include
   a. the agents of transmission
   b. the number of invaders
   c. the virulence of the organisms
   d. the portal of entry
   *e. all of the above
2. A disease that is very easily transferred from one person to another is
   a. communicable
   *b. contagious

   c. latent
   d. chronic
   e. cancer
3. The period between the time when an organism finds a host and the host manifests the illness is the
   *a. incubation period
   b. illness period
   c. convalescent period
4. Endotoxins are
   a. toxic substance excreted from the bacterial cell
   b. fluids, often from formed elements of blood
   *c. large molecules of lipopolysaccharide that are normal components of gram-negative bacterial cell walls
   d. exotoxin released in food by certain staphylococci
5. A capsule increases the virulence of certain pathogenic bacteria by
   *a. aiding the cell to evade phagocytosis
   b. destroying white blood cells
   c. inhibiting the formation of protective substances
   d. destroying red blood cells
6. The ability of an organism to produce a capsule affects
   *a. its invasiveness
   b. its virulence
   c. the way it is transferred
   d. all of the above
   e. none of the above
7. The ability of organisms to produce disease depends on the production of
   a. exotoxins
   b. enzymes
   c. capsules
   d. endotoxins
   *e. any of the above
8. An infection in one body area that spreads to other sites is referred to as a (an)
   a. local infection
   *b. focal infection
   c. inapparent infection
   d. fulminating infection
9. One can establish with certainty that a particular organism is the causative agent of a given disease on the basis of
   a. a culture made from the infected area
   b. a stain of a direct smear from the infected areas
   *c. Koch's postulates
   d. observation of the clinical picture
10. The invasiveness of *Streptococcus pneumoniae* is increased because the capsule
    *a. interferes with phagocytosis
    b. stimulates rapid growth

c. produces a toxin
11. Symptoms of the inflammatory process are
    a. swelling and pain
    b. redness and heat
    c. loss of function of the affected area
    *d. *a* and *b*
    e. all of the above
12. An enzyme that dissolves a fibrin clot is
    a. coagulase
    b. hyaluronic acid
    *c. streptokinase
    d. hemolysin
    e. none of the above
13. During an inflammation response, pyogenic bacteria are responsible for
    a. redness
    b. swelling
    c. heat
    *d. pus formation
14. A disease that is transmitted from one person to another is called
    a. endemic
    *b. communicable
    c. sporadic
    d. contagious
    e. pandemic
15. Any agent that carries a disease from one person to another is called a (an)
    a. episome
    *b. vector
    c. host
    d. parasite
    e. FBI
16. An example of a disease that is transmitted by direct intimate contact is
    a. typhoid
    *b. syphilis
    c. dysentery
    d. cholera
    e. poliomyelitis
17. An example of a disease that enters the body through the gastrointestinal tract via the mouth is
    a. measles
    b. gonorrhea
    *c. infectious hepatitis
    d. malaria
    e. yellow fever
18. A disease constant in a certain locality is said to be
    a. epidemic
    *b. endemic
    c. worldwide
    d. sporadic
    e. pandemic
19. All of the following are significant factors in the development of disease *except*

a. portal of exit from host
b. portal of entry
*c. interval time
d. number of invaders
e. virulence of organisms
20. The common cold is usually
    a. epidemic
    b. pandemic
    c. sporadic
    *d. endemic
    e. septicemic
21. Which of the following diseases is typically transmitted through air?
    a. sleeping sickness
    b. dysentery
    *c. common cold
    d. smallpox
    e. amebiasis
22. The primary danger in water lies in pathogens spread by
    *a. fecal contamination
    b. airborne contamination
    c. soil contamination
    d. respiratory secretions
23. The source of salmonellae in food infection is
    *a. the intestinal tract of many animals
    b. only human fecal contamination
    c. soil
    d. nose and throat
24. The common sources of *Staphylococcus aureus* are
    a. soil
    b. water
    c. the nose and throat
    d. the hands
    *e. only *c* and *d*
25. A disease that occurs in epidemic proportions throughout the world is said to be
    a. endemic
    b. epidemic
    *c. pandemic
    d. sporadic
26. Gonorrhea and syphilis are transmitted by
    a. person-to-person contact
    *b. mucous-to-mucous contact
    c. food, water, and soil
    d. direct skin contact
27. Which group(s) constantly strive(s) to prevent epidemics?
    a. State governments
    b. The World Health Organization
    c. The U.S. Public Health Service
    *d. All of the above
    e. *b* and *c* only
28. When unsure about the purity of drinking water, one should
    a. freeze water and thaw quickly

b. heat water until warm and let cool
c. warm water for 20 minutes
d. do nothing; bacteria in water are rarely pathogenic
*e. boil water for 5 minutes

29. Items such as bedclothes, towels, and sheets, which transfer pathogens to a susceptible host, are called
    a. vectors
    b. dangerous
    *c. fomites
    d. contaminated

30. Any site in which a pathogen can survive and multiply until it is transferred to a host is known as a (an)
    a. infection
    *b. reservoir
    c. inflammation

Chapter 8
Preventing the Spread of Communicable Diseases

Objectives

1. List the six factors that have contributed to an increase in hospital-acquired (nosocomial) infections
2. List areas in the hospital where nosocomial infections are most probable
3. List several types of patients who are extremely vulnerable to infectious diseases
4. Write a brief description of reverse isolation and source isolation
5. Briefly describe the important procedures to follow in universal precautions
6. Discuss the role of the health worker in the collection of specimens
7. List the types of specimens that usually must be collected from patients
8. Discuss the general precautions that must be observed during the collection and handling of specimens
9. Describe the proper procedure for obtaining specimens
10. Discuss the importance of quality control in a microbiology laboratory
11. List the sources of water contamination
12. Describe how water and sewage are treated
13. Discuss how epidemics are controlled and prevented

Summary and Discussion

It is especially important for those studying to become health workers in clinics and hospitals to understand the basic techniques used to pre-vent the spread of diseases from infectious patients to susceptible patients and themselves. The patients most vulnerable to hospital infections are newborns and sick or incapacitated patients, including women in delivery; surgical, diabetic, and cancer patients; and paralyzed individuals. The students should also be aware of the roles played by the hospital laboratory and epidemiologists from the national, state, and local public health departments.

Students should think of pathogens in the clinic and hospital as large glowing objects to be handled very carefully. Thus, they are prepared to learn to collect, handle, process, and test specimens properly.

Outline

I. Prevention of hospital infections
  A. How hospital infections develop
    1. Hospital-acquired and community-acquired infections
    2. Protection of patients
    3. Protection of health care workers
  B. General control measures
  C. Infection control procedures
    1. Medical and surgical asepsis
    2. Universal precautions
    3. Patient isolation
  D. Hospital infection control
  E. Medical waste disposal
II. Specimen collection, processing and testing
  A. Role of health care personnel
  B. Proper collection of specimens
  C. Shipping procedures
  D. Identification and drug sensitivity
  E. Quality control
III. Environmental disease control measures
  A. Public health authorities
  B. Water supplies and sewage disposal
    1. Water pollution
    2. Water and sewage treatment

Suggested Laboratory Exercises

1. Press a ring, necklace, or other piece of jewelry onto an agar plate. Remove the jewelry. Press a hair onto the other half of the agar. Incubate the plate 24 hours and observe the number of colonies.
2. Dip the tip of a thermometer that has been disinfected into a tube of nutrient broth. Incubate the broth for 24 hours and observe. Cloudiness indicates the growth of microorganisms. Streak a loopful of the cloudy broth onto an agar plate, incubate for 24 hours, and count the number of

colonies. Note that each colony arises from one microorganism.

Suggested Audiovisual Aids

1. *The Fight Against Infectious Diseases.* 16-mm, color, sound, 21 min. (CIBA)
2. *AIDS: Our Worst Fears.* Color videotape, QB–1052, 57 min. (FFH)
3. *Death of a Disease.* A NOVA color videotape, 58 min. (TLV)
4. *Airborne Transmission of Tubercle Bacilli.* 16-mm, color, sound, 7 min. (NMAVC)
5. *Asepsis.* 16-mm, color, sound, 15 min. (CU)
6. *Microbial Sampling of the Operating Room Environment.* 16-mm, color, sound, 22 min. (D&G)
7. *Sputum Collection and Handling.* 16-mm, color, sound, 7 min. (NAC)
8. *Urine Collection.* 16-mm, color, sound, 16-min. (EAT)

Sample Examination Questions

1. In today's hospitals most infections are spread by
   a. air
   b. food
   *c. lack of medical asepsis
   d. body excreta
   e. direct patient contact
2. Medical asepsis aims at
   a. elimination of all forms of microbial life in a given area
   *b. cleanliness and control of disease agents
   c. establishing hand-washing techniques.
3. Maintenance of surgical asepsis is
   a. an easy task
   b. of no real concern
   *c. a most difficult and exacting task
   d. just a general procedure
4. Surgical asepsis
   a. is the same as medical asepsis
   *b. comes into play when the primary barrier (the skin) is broken
   c. part of the daily routine of every individual in a hospital
5. Surgical aseptic technique includes all of the following *except*
   a. wearing a sterile gown
   b. use of a mask
   c. use of a cap
   d. careful and thorough washing of hands
   *e. touching a wound with bare hands
6. Surgical asepsis aims at

   a. dust control
   b. destroying pathogens
   c. cleanliness and control of disease agents
   *d. elimination of all forms of microbial life in a given area
   e. reducing the number of microorganisms in the environment
7. A carrier may be a person who
   a. has an unrecognized infection with no symptoms of illness
   b. has recovered from a disease
   c. has no symptoms of an illness but harbors a pathogenic organism
   *d. all of above
8. Preferably, instruments used in surgical procedures should be sterilized by
   a. toxic chemicals
   b. soaking
   c. sunlight
   d. ultraviolet light
   *e. autoclaving
9. Soiled dressings should be
   *a. wrapped, autoclaved, and discarded
   b. soaked in disinfectant
   c. flushed down the toilet
   d. thrown in the wastebasket
10. General disease control measures in the kitchen include
   a. personal cleanliness
   b. sanitary handling of food with careful washing
   c. thorough cooking
   d. proper refrigeration
   *e. all of the above
11. The major cause of wound infections resulting in gas gangrene is
   a. *Salmonella*
   *b. *Clostridium perfringens*
   c. *Clostridium botulinum*
   d. *Staphylococcus*
12. The indicator organism for fecal pollution in the United States is
   a. *Streptococcus faecalis*
   b. *Clostridium perfringens*
   *c. *Escherichia coli*
   d. *Clostridium tetani*
13. The disinfection process is aimed at
   a. killing all nonpathogens
   *b. destruction of disease agents
   c. killing all bacteria present
14. Sterilization may be defined as
   a. destruction of most organisms
   b. killing all vegetative cells but not spores
   *c. killing all forms of life in a given area
   d. physical removal of bacteria
15. The most usual and efficient method of sterilization for most objects is

*a. heat
b. cold
c. radiations

16. After 10 minutes in boiling water, *Escherichia coli* was killed, but *Bacillus subtilis* was found to survive. Which of the following provides the most probable explanation?
   a. *B. subtilis* forms a rigid cell wall
   *b. *B. subtilis* forms endospores
   c. *E. coli* was actively growing
   d. *B. subtilis* forms capsules

17. Protective isolation can also be called
   *a. reverse isolation
   b. enteric precautions
   c. respiratory isolation
   d. wound and skin precautions
   e. none of the above

18. In the hospital, the patients most vulnerable to infections or diseases are
   *a. burn patients
   b. patients with allergies
   c. young adults
   d. cancer patients
   e. *b* and *d* only

19. The most prevalent opportunistic bacteria found in hospitals are
   a. *Staphylococcus aureus*
   b. *Escherichia coli*
   c. *Streptococcus pneumoniae*
   *d. *Pseudomonas* and *Proteus* species
   e. all of the above

Chapter 9
Human Defenses Against Disease

Objectives

1. List and describe the nonspecific defenses of the human body, or the first line of defense
2. Define phagocytosis, the body's second line of defense
3. List the various types of phagocytic cells
4. Describe the process of inflammation
5. Describe the immune response, or "third line of defense"
6. Define *antigen, antibody,* and *immunoglobulin*
7. Differentiate between active and passive immunity
8. Compare natural and artificial immunity
9. List three ways in which vaccines are prepared
10. Draw a graph representing the primary and secondary antibody responses to antigens
11. Differentiate between immediate and delayed hypersensitivity

12. Define autoimmunity and give examples
13. List five serological tests used to determine the presence of a specific antibody in human serum

Summary and Discussion

The body has many mechanisms of defense even without the immune system. The first line of defense includes the mechanical barriers of the skin and mucous membranes and chemical substances secreted by these tissues and cells. The second line of defense includes phagocytosis and the inflammatory response. The third line of defense is the immune response, or the production of antibodies.

Immunology is a very complex and difficult aspect of microbiology. However, it is important to understand the general concept of the normal immune response and the types of immunity that can be acquired. It is often difficult to differentiate the normal protective immune response from the harmful aspects of hypersensitivities, such as allergies, anaphylactic shock, and autoimmunity. There are many serologic tests developed by research and clinical laboratories to determine if a patient is susceptible to a disease, immune to a disease, has a clinical case of a disease, or is hypersensitive to a substance, by indicating the presence of antibody in the serum or other areas. The student should understand the concept involved in some of these tests.

Outline

I. First line of defense: mechanical and chemical barriers
   A. Skin
      1. Perspiration
      2. Oily secretion
      3. Normal flora
   B. Mucous membranes of body openings
      1. Mucous traps
      2. Flushing action
      3. Ciliary action
      4. Lysosomes
      5. Digestive enzyme
      6. Normal flora
II. Second line of defense
   A. Interferon
   B. Interleukins
   C. Complement
   D. Properdin
   E. Prostaglandins
   F. Phagocytosis
   G. Inflammation
III. Immunity, third line of defense

A. Innate (nonspecific) resistance
B. Acquired immunity
  1. Active
     a. Naturally acquired through disease
     b. Artificially acquired through vaccine
  2. Passive
     a. Naturally acquired (congenital)
     b. Artificially acquired through antiserum, gamma globulin
IV. Immunology
  A. Antigens
  B. Antibodies, immunoglobulins
V. The immune system
VI. The immune response
  A. Antibody production
  B. Antibody structure and function
     1. Immunoglobulin G: serum and lymph
     2. Immunoglobulin M: serum macroglobulins
     3. Immunoglobulin A: secretions
     4. Immunoglobulin E: tissue bound
     5. Immunoglobulin D
  C. Hypersensitivity and cell-mediated immune responses
     1. Immediate
        a. The allergic response (type I)
        b. Cytotoxic hypersensitivity (type II)
        c. Immune complex reactions (type III)
           (1) Serum sickness
           (2) Autoimmune diseases
     2. Delayed: cell-mediated immunity (type IV)
        a. Transplantation rejection
        b. Tuberculin and fungal skin tests
        c. Contact dermatitis
VII. Immunodiagnostic procedures

Suggested Laboratory Exercises

1. Cough forcefully onto a blood agar plate held a few inches from the mouth and again onto another held 10 or more inches from the mouth. Incubate the plates at 37°C for 24 hours and examine the colonies.
2. Examine a stained smear of pus containing phagocytes. What is a pus cell and what is its function?
3. *Demonstrations.* The student or the instructor may perform many demonstrations illustrating agglutination, precipitation, hemolysis, toxin

neutralization, opsonization, and capsular swelling tests by using various commercially prepared antigens and corresponding antibodies.

Suggested Audiovisual Aids

1. *The Body Against Disease.* Color videotape, W-730-VS (HC)
2. *Phagocytosis and Degranulation.* 16-mm, black and white, sound, 14 min. (AFIP)
3. *Infectious Diseases and Natural Body Defenses.* 16-mm, color, sound, 11 min. (COR)
4. *Cellular and Molecular Aspects of the Immune Response.* 16-mm, black and white, sound, 43 min. (NMAVC)
5. *A Question of Immunity.* 16-mm, color, sound, 13 min. (HAF)
6. *Human Immune System: The Fighting Edge.* Color videotape, QB-1558, 52 min. (FFH)
7. *Autoimmunity and Disease.* 16-mm, color, sound, 33 min. (MSD)
8. *Rabies: F-A Staining.* 16-mm, color, sound, 8 min. (NMAVC)

Sample Examination Questions

1. Interferon
   a. is cell-specific, not virus-specific
   b. is a small protein molecule
   c. can inhibit certain viral infections
   *d. all of the above
   e. none of the above
2. Interferon is
   a. an antiviral substance
   b. virus-specific
   c. cell-specific
   d. all of the above
   *e. *a* and *c*
3. What stimulates the leukocytes to migrate to an injured area of the body?
   a. Phagocytosis
   *b. Chemotaxis
   c. They just do it
   d. Leukocytes do not migrate to an injured area
   e. Complement
4. The nonspecific defense mechanisms of the skin include
   a. oily secretions
   b. a mechanical barrier
   c. acidic perspiration
   *d. all of the above
   e. *a* and *b*
5. The mucous membranes are part of the body's defenses because they

a. contain enzymes
b. have a flushing action
c. serve as traps
*d. all of the above
e. none of the above

6. Indigenous microflora are found in all the following areas *except*
   a. the mouth
   b. the anus
   *c. the bladder
   d. the vagina
   e. the intestine

7. One of our major safeguards against infection is
   *a. the unbroken skin
   b. hand-washing
   c. the teeth
   d. clothing
   e. body hair

8. Which is considered the first line of defense?
   *a. The skin
   b. Interferon
   c. Phagocytes
   d. Serum proteins
   e. Properdin

9. Which of the following specifically destroys the cell walls of bacteria?
   a. Interferon
   *b. Lysozyme
   c. Phagocytes
   d. Serum proteins
   e. Properdin

10. The second line of defense includes
    a. phagocytes
    b. serum proteins
    c. interferon
    d. *b* and *c* only
    *e. all of the above

11. The third line of defense is
    a. the cells
    b. lysozymes
    *c. specific antibodies
    d. nonspecific
    e. none of these

12. Innate or natural immunity is inborn and
    a. is independent of previous experience
    b. often depends on the activities of phagocytes
    c. is a result of the genetic constitution of an individual
    *d. all of the above

13. Naturally acquired active immunity
    a. occurs following natural exposure to a foreign agent
    b. occurs following exposure to viruses
    c. includes antibody production
    d. can be long-lasting or temporary
    *e. all of the above

14. Naturally acquired passive immunity
    a. occurs following administration of protective antibodies
    b. occurs following immunization with, for example, polio virus
    *c. includes placental transfer of immunity to poliomyelitis from mother to fetus
    d. occurs after a case of poliomyelitis

15. Artificially acquired active immunity
    a. occurs following immunization
    b. includes the polio virus vaccine
    c. occurs following administration of protective antibodies
    *d. *a* and *b*
    e. all of the above

16. Artificially acquired passive immunity
    a. is acquired by administration of protective antibodies
    b. includes placental transfer of preformed antibodies against polio virus
    c. includes placental transfer of immunity to poliomyelitis
    d. includes immunity of man to distemper viruses of cats and dogs
    *e. *a* and *b*

17. Antibodies
    a. are proteins
    b. have a molecular weight of 160,000 to 900,000
    c. are immunoglobulins
    d. are found in the gamma globulin portion of serum when blood proteins are separated
    *e. all of the above

18. Antigens may be
    a. proteins
    b. polysaccharides
    c. nucleic acids
    d. *a* and *b*
    *e. all of the above

19. Which of the following is an important characteristic of antigens?
    a. They have low molecular weights.
    b. They are monospecific.
    *c. They are usually foreign to the host that forms the specific antibodies to them.
    d. They are bivalent.

20. Five major classes of human immunoglobulins have been distinguished. Which one is responsible for certain allergies?
    a. IgM
    b. IgG
    c. IgD
    *d. IgE
    e. IgA

21. Which class of immunoglobulins is the most abundant in the human body?
    *a. IgG
    b. IgA
    c. IgM
    d. IgE
22. Which of the following is *not* a characteristic of IgG?
    a. It functions for protection within the body
    b. It protects mucous membranes
    *c. It attaches to mast cells and basophils
    d. It attaches to phagocytes
    e. It combines with a specific antigen
23. Haptens
    a. can react with specific antibodies
    b. cannot incite the production of antibodies unless chemically combined with large carrier molecules
    c. may be antigenic determinants
    d. have low molecular weights
    *e. all of the above
24. When an individual responds immunologically against his own body constituents, this is called
    a. anamnestic response
    b. immunocompetency
    c. humoral immunity
    *d. autoimmunity
25. Antibody molecules are produced in lymphoid tissues by
    a. B lymphocytes
    b. macrophages
    c. plasma cells
    d. basophils
    *e. *a* and *c*
    f. all of the above
26. Which cells synthesize and secrete humoral antibodies?
    *a. Plasma cells
    b. T Lymphocytes
    c. Monocytes
    d. Leukocytes
27. The titer of a serum indicates
    a. the amount of antibodies present
    b. the presence of clotting factors
    c. the presence of antibodies to a disease
    d. antibodies present in serum
    *e. *a, c* and *d*
28. Which of the following is true of both precipitin and agglutination reactions?
    a. The patient's serum must first be heated.
    b. They both use soluble antigens.
    c. Extremely large aggregates of antigen and antibody are formed.
    *d. They both aid in the removal of antigens from the circulation.
29. Cell-mediated immunity

    a. involves both T lymphocytes and macrophages
    b. can be passively transferred only with sensitized cells
    c. is involved in tuberculosis
    d. is of major importance in the response to most tumors and to foreign cells in tissue transplants
    *e. all of the above
30. The anamnestic response
    a. is also called the memory response
    b. designates the events that occur when the antigen is contacted for the second or subsequent time
    c. is more intense than the primary response
    d. involves the survival of immunocompetent cells from the first exposure as memory cells
    *e. all of the above
31. In complement-fixation reactions
    a. bacteria or other cells may undergo lysis
    b. the concentration of antibodies may be measured
    c. precipitation occurs
    d. the antigen-antibody complex must first be present
    e. all but c
32. The neutralizing substance developed by an animal upon injection of a toxoid is called
    a. immunity
    b. vaccine
    *c. antitoxin
    d. complement
33. The type of response connected with blood transfusion reaction is
    a. IgE mediated
    b. IgG mediated
    *c. cytotoxic
    d. immune complex
34. Specifically sensitized lymphoid cells are responsible for
    a. hay fever reaction
    b. serum sickness
    c. anaphylactic shock
    *d. positive tuberculin skin test result
35. The nonantigenic antibiotic penicillin can cause an allergic response because
    *a. it becomes a hapten-carrier complex
    b. it is a very antigenic drug
    c. it attacks platelets
    d. it settles in the heart
36. The reason why IgE antibodies cause hay fever and its symptoms is because
    *a. the mast cells they attach to are abundant in the nose and lungs
    b. IgE is found in ragweed pollen

c. IgE secretes mucus

d. IgE initiates sneezing

37. A *common* way of "desensitizing" a person who has a specific allergies is
   a. removing his antibodies
   b. desensitizing his mast cells
   *c. giving multiple small injections of the specific antigen
   d. moving him to another geographic area

38. Blood typing is a good example of what kind of test?
   a. Precipitation
   b. Opsonization
   *c. Agglutination
   d. Immobilization

39. The secretory antibody found in tears, saliva, and colostrum is
   a. IgG
   b. IgE
   *c. IgA
   d. IgM
   e. IgB

40. An antigen that causes an allergic reaction is an
   *a. allergen
   b. antibody
   c. antitoxin
   d. antibiotic
   e. *c* and *d*

41. The factor(s) involved in the development of hypersensitivity is (are)
   a. the nature and amount of antigen
   b. the length of time and frequency of exposure to antigen
   c. ability of the body to produce IgE antibodies
   *d. all of these

42. The first immunoglobulin that is formed against infectious agents is
   a. IgG
   *b. IgM
   c. IgA
   d. IgE
   e. IgD

43. Persons unable to produce antibodies have an abnormality called
   a. nonprimary response
   *b. agammaglobulinemia
   c. hypogammaglobulinemia
   d. hyperglobulinemia
   e. none of the above

Chapter 10
Major Diseases of the Body Systems

Objectives

1. Name the major organs that might become infected in each body system
2. List the most common normal microbial flora usually found in the various body systems
3. Outline the causative agent, reservoir, mode of transmission, pathogenesis, treatment, and control measures for the major infectious diseases of each system
4. For each body system, list some examples of diseases that are caused by bacteria, viruses, fungi, or protozoa

Summary and Discussion

A survey of the major infectious diseases of each body system is presented in this chapter. A drawing of each normal body system provides the anatomy for discussion purposes. The indigenous microflora and nonspecific defense mechanisms of each area are discussed because the source of the pathogen may be opportunistic indigenous microflora in the compromised host.

Each infectious disease is presented in outline format, indicating characteristics of the disease, mechanisms of pathogenicity, reservoirs, modes of transmission, incubation periods, means of control, and usual treatments.

This material can best be presented by discussing the anatomy, indigenous microflora, defense mechanisms, preventive measures, and then the diseases of each area. Most diseases are outlined in the body system where they cause the most damage or where they enter the body; however, many of these infections may move from one area to another, involving several body systems.

Outline

I. Diseases of the skin
   A. Viral infections
      1. Chickenpox
      2. German measles, rubella
      3. Measles, rubeola
      4. Warts
   B. Bacterial infections
      1. Impetigo
      2. Scarlet fever, scarlatina
      3. Boils, carbuncles
      4. Acne

5. Anthrax
6. Leprosy
C. Fungal infections
1. Dermatomycosis, tinea, ringworm
D. Burn and wound infections
II. Diseases of the eye
A. Viral infections
1. Viral conjunctivitis
B. Bacterial infections
1. Bacterial conjunctivitis, pink eye
2. Chlamydial conjunctivitis, paratrachoma
3. Trachoma
4. Gonococcal conjunctivitis
III. Diseases of the mouth
A. Bacterial infections
1. Dental caries, gingivitis, peridontitis
2. Acute necrotizing ulcerative gingivitis (ANUG), trench mouth, Vincent's angina
IV. Diseases of the ear
A. Bacterial infections
1. Otitis media, middle-ear infections
2. External otitis, swimmer's ear
V. Diseases of the respiratory system
A. Nonspecific infections
1. Pneumonia
B. Viral infections
1. Common cold
2. Croup
3. Influenza
C. Bacterial infections
1. Diphtheria
2. Legionellosis
3. Strep throat
4. Tuberculosis
5. Whooping cough
D. Protozoan infections
1. Pneumocystis pneumonia
VI. Diseases of the gastrointestinal tract
A. Viral infections
1. Gastroenteritis
2. Hepatitis A
3. Hepatitis B
B. Bacterial infections
1. Campylobacter gastroenteritis
2. Cholera
3. Enteropathogenic E. coli diarrhea
4. Salmonellosis, typhoid fever
5. Shigellosis
C. Foodborne intoxications, food poisoning
1. Botulism
2. Clostridium perfringens food poisoning
3. Staphylococcal food poisoning
D. Protozoan diseases

1. Amebiasis
2. Giardiasis
VII. Diseases of the urogenital tracts
A. Sexually transmitted diseases
B. Urinary tract infections
1. Urethritis, cystitis, ureteritis
C. Viral infections
1. Genital herpes
2. Genital warts
D. Bacterial infections
1. Chlamydial genital infections
2. Gonorrhea
3. Syphilis
E. Protozoan infections
1. Trichomoniasis
F. Other STDs
VIII. Diseases of the circulatory system
A. Viral infections
1. AIDS
2. Colorado tick fever
3. Infectious mononucleosis
4. Mumps
B. Rickettsial infections
1. Rocky Mountain spotted fever
2. Endemic flea-borne typhus fever
3. Epidemic louse-borne typhus fever
C. Bacterial infections
1. Subacute bacterial endocarditis (SBE)
2. Plague
3. Tularemia
D. Protozoan infections
1. Malaria
2. Toxoplasmosis
3. Trypanosomiasis
IX. Diseases of the nervous system
A. Viral infections
1. Poliomyelitis
2. Rabies
B. Bacterial infections
1. Tetanus

Suggested Laboratory Exercises

1. Perform a hand-washing experiment by touching the unwashed fingers to half of an agar plate. Then scrub hands and fingers with a strong soap and brush, dry the hand, and touch fingers of the same hand to the other half of the nutrient agar plate. After incubating 24 hours at 37°C, many colonies will be observed; perhaps more colonies will be seen from the scrubbed fingers showing the indigenous microflora that live deep within the epidermal layers of the skin.
2. With a toothpick, remove debris from

outer gingival sulcus; streak carefully (without breaking the agar) on a nutrient or blood agar plate. Flame loop and streak the gum inoculum over the plate for isolated colonies. Incubate at 37°C and observe.

Suggested Audiovisual Aids

1. *Death of a Disease.* A NOVA color videotape, 58 min. (TLV)
2. *Common Cold.* Color videotape, XC–1801, 28 min. (FFH)
3. *Hunt for the Legion Killer.* Color videotape, 57 min. (TLV)
4. *Hepatitis.* Color videotape, DFMD–107, 29 min. (PBS)
5. *Lyme Disease.* Color videotape, XC–1714, 26 min. (FFH)
6. *The Intimate Epidemic.* Color videotape, EB–2037, 24 min. (FFH)

Sample Examination Questions

1. Which one of the following diseases is caused by a protozoan?
   a. Syphilis
   b. LGV
   c. NGU
   d. Herpes
   *e. None of these
2. Which of the following diseases is caused by a virus?
   a. dental caries
   *b. influenza
   c. tetanus
   d. amebiasis
   e. candidiasis
3. Which of the following diseases is caused by fungi?
   a. dental caries
   b. influenza
   c. tetanus
   d. amebiasis
   *e. candidiasis
4. Fungi causing systemic mycoses include
   a. *Blastomyces dermatitidis*
   b. *Histoplasma capsulatum*
   c. *Coccidioides immitis*
   *d. all of the above
5. Group D, α-hemolytic streptococci are involved in
   a. dental caries
   b. pneumonia
   *c. subacute bacterial endocarditis
   d. *a* and *c* only
   e. all of the above

6. Pneumococci can be subdivided into approximately 100 types based on
   a. flagellar antigen
   *b. capsular antigen
   c. somatic antigen
   d. *a* and *c* only
   e. all of the above
7. A frequent cause of meningitis is
   a. *Haemophilus influenzae*
   b. *Streptococcus pneumoniae*
   c. *Neisseria meningitidis*
   *d. all of the above
8. The DPT vaccine, which is given to young infants and children, represents protection against
   a. Diplococcus, pertussis, tetanus
   *b. Diphtheria, pertussis, tetanus
   c. Diphtheria, pneumococcus, tetanus
   d. Diphtheria, pertussis, treponema
9. The procedure used to identify *Mycobacterium* is
   a. Shick test
   b. Schultz-Charlton reaction
   *c. Acid-fast stain
   d. Dick test
10. The genus *Streptococcus* can be characterized as
    a. spherical cells
    b. growing in chains
    c. gram-positive
    d. *a* and *c* only
    *e. all of the above
11. Complete clearing of the media around a *Streptococcus pyogenes*, group A, colony is described as
    a. alpha hemolysis
    *b. beta hemolysis
    c. gamma hemolysis
    d. delta hemolysis
12. From the symptoms listed, identify those associated with a "strep throat."
    A. Inflamed mucous membranes; B. Purulent exudate; C. Low to moderate temperature; D. Spread through school contacts.
    a. A, B, C, D
    b. A, C, D
    *c. A, B, D
    d. A, B, C
    e. B, C, D
13. Which test is used to identify group A streptococci?
    a. Bacitracin sensitivity
    b. Neufeld/quellung test
    c. Fluorescent antibody techniques
    *d. *a* and *c* only
    e. All of the above
14. Pericarditis is a disease of the
    a. lungs

b. bronchi
*c. heart
d. two of these
e. all of these

15. Infectious mononucleosis is caused by
    a. *Haemophilus aegyptius*
    b. *Haemophilus influenzae*
    c. *Escherichia coli*
    d. *Chlamydia trachomatis*
    *e. Epstein-Barr virus

16. Tinea pedis refers to
    a. the species name of fungus
    *b. disease of the feet
    c. name of a flea vector of disease
    d. species of normal skin bacterium

17. Tears contain an antibacterial chemical which is
    a. sebum
    *b. lysozyme
    c. complement
    d. IgG
    e. lymphokines

18. *Streptococcus mutans* is of concern in caries because
    a. it forms dextrans
    b. plaque contains this bacterium
    c. it ferments sugar and produces acids
    d. it is commonly found in the mouth
    *e. all of the above

19. Dental caries are most likely to occur when which of the following factors are involved?
    a. High dietary sugar, poor hygiene, no acidogenic bacteria
    b. High dietary sugar, good hygiene, no acidogenic bacteria
    *c. High dietary sugar, poor oral hygiene, acidogenic bacteria
    d. Low dietary sugar, poor oral hygiene, acidogenic bacteria

20. The eustachian tube is normally associated with which function?
    a. Drains tears from the eye to pharynx
    b. Air passageway in cranial bones near nares
    *c. Equilibrium of air pressure on eardrum
    d. Producing saliva

21. Otitis media is primarily an infection of the
    *a. middle ear
    b. parotid salivary gland
    c. throat
    d. sinuses

e. pharynx

22. Primary atypical pneumonia is caused by a
    a. gram-positive coccus
    b. gram-positive rod
    c. gram-negative rod
    *d. mycoplasma

23. The common cold is caused by
    a. adenoviruses
    b. coronaviruses
    c. rhinoviruses
    d. echoviruses
    *e. all of the above

24. The major difficulty in preparing an effective vaccine against the common cold is that
    a. the cold is an autoimmune disease
    *b. there are so many different causative agents that it is impossible to immunize a person against all of them
    c. all vaccines prepared thus far have been too highly allergenic to be used safely
    d. the cold is a viral disease, and immunization against viral disease is generally ineffective
    e. the causative agent cannot be attenuated

25. Cases of diphtheria in the community occur because
    a. some people are not immunized
    b. elderly persons have lost their immunity
    c. healthy carriers of the bacteria are present
    *d. all of the above

26. NANB refers to
    a. a skin test for botulism
    b. a vaccine for *Clostridium botulinum* types A and B
    *c. non-A, non-B hepatitis virus
    d. neurotoxin produced by dinoflagellates

27. Cystitis symptoms are caused by
    *a. infections of the bladder and urethra
    b. infections of the kidney
    c. infections of the uterus
    d. normal flora of the bladder

28. Place the following phases of syphilis infection in proper order.
    A. Rash; B. Chancre; C. Brain infection.
    a. A, B, C
    *b. B, A, C,
    c. C, A, B
    d. A, C, B

Sources for Audiovisual Aids

The sources listed below correspond to a specific abbreviation. These abbreviations are used in the lists of audiovisual materials pertinent to each chapter.

AFIP
Armed Forces Institute of Pathology
Audiovisual Support Center
Fort Sam Houston, TX 78234

ASM
American Society for Microbiology
1325 Massachusetts Ave. NW
Washington DC 20005-4171

ASF
Association Films, Inc.
866 Third Avenue
New York, NY 10022

BYU
Brigham Young University
Media Marketing
W–STAD
Provo, UT 84602

CIBA
Ciba Pharmaceutical Co.
566 Mooris Avenue
Summit, NJ 07901

CM
Concept Media
1500 Adams Avenue
Costa Mesa, CA 92626

COR
Coronet/MTI Film and Video
108 Wilmot Road
Deerfield, IL 60015

CU
University of Colorado
Educational Media Center
Boulder, CO 80309

D & G
Davis and Geck Distributors
American College of Surgeons
Surgical Film Library
1 Casper Street
Danbury, CT 06810

EAT
Eaton Medical Film Library
Eaton Laboratories
Morton-Norwich Products, Inc.
13–27 Eaton Avenue
Norwich, NY 13815

EB
Encyclopedia Britannica Educational
    Corporation
301 South Michigan Avenue
Chicago, IL 60604

EI
Educational Images
P.O. Box 3456
West Side
Elmira, NY 14905

EL
Eli Lilly & Co.
Educational Resources Program
P.O. Box 100B
Indianapolis, IN 46206

EU
Emory University
Audiovisual Librarian
A.W. Calhoun Medical Library
Woodruff Memorial Building
Atlanta, GA 30322

FFH
Films for Humanities
743 Alexander Rd.
Princeton, NJ 08540

HC
Harper & Row Media
Harper & Row Publishers, Inc.
Distributed by MTI Teleprograms
108 Wilmot Road
Deerfield, IL 60015

HAF
HAF Alternatives on Film
P.O. Box 22141
San Francisco, CA 94122

IU
Indiana University
Audiovisual Center
Bloomington, IN 47405

LL
Lederle Laboratories Division
American Cyanamide Co.
Pearl River, NY 10965

MG
McGraw-Hill Book Co.
330 West 42nd Street
New York, NY 10036

MMP
Multi-Media Publishing, Inc.
1393 South Inca Street
Denver, CO 80223

MSD
Mercke Sharp and Dohme
Health Information Services Library
West Point, PA 19486

NAC
National Audio-Visual Center (CSA)
Rental Branch or Sales Branch
Washington, DC 20409

NEM
National Educational Media, Inc.
21601 Devonshire Street, #300
Chatsworth, CA 91311

NHF
National Health Films
P.O. Box 13973, Station K
Atlanta, GA 30324

NMAVC
National Medical Audiovisual Center
2111 Plaster Bridge Road
Atlanta, GA 30324

PBS
PBS Video
1320 Braddock Place
Alexandria, VA 22314

PD&C
Parke, Davis and Company
Box 118, General Post Office
Detroit, MI 48232

PLD
Pfizer Laboratories Division
Film Library
470 Park Avenue South
New York, NY 10016

PLP
Point Lobos Productions
20417 Califa Street
Woodland Hills, CA 91367

SPFL
Schering Professional Film Library
Southern Film Distribution Center
c/o Association Films, Inc.
8615 Directors Row
Davis, TX 75247

SQ
E.R. Squibb & Sons
Lawrenceville-Princeton Road
Princeton, NJ 08540

STAN
Stanton Films
2417 Artesia Boulevard
Redondo Beach, CA 90278

SEF
Sterling Educational Films
241 East 34th Street
New York, NY 10016

SLUMCL
St. Louis University Medical Center
    Library
1402 South Grand Boulevard
St. Louis, MO 63104

TFI
Teaching Films, Inc.
Division of Audiovisual Corp.
930 Pitner Avenue
Avanston, IL 60202

TLV
Time-Life Film and Video
100 Eisenhower Drive
Paramus, NJ 07652

TS
Telestar Productions
366 N. Prior Ave.
St. Paul, MN 55104

UNITED
United World Films, Inc.
221 Park Avenue South
New York, NY 10003

UPFL
Upjohn Professional Film Library
7000 Portage Road
Kalamazoo, MI 49001

WSU
Wayne State University
Film Library
77 West Canfield Avenue
Detroit, MI 48202

WYA
Wyeth-Ayerst Laboratories
P.O. Box 8299
Philadelphia, PA 19101–8299

Suggestions for Final Examination

Questions may be drawn from previously given tests, student workbook matching and true-false questions, as well as some new questions on the general concepts of microbiology. The following are suggested final examination questions.

Multiple Choice

Choose the best answer for each of the following questions.
1. In which of the following genetic recombinations does the episome consist of phage DNA?
   a. Transformation
   b. Transduction
   *c. Phage conversion
   d. Conjugation
2. Which of the following is *not* a viral disease?
   a. Rabies
   *b. Gonorrhea
   c. Mumps
   d. Measles
   e. Smallpox
3. Which of the following is *not* a fungal disease?
   *a. Amebiasis
   b. Athlete's foot
   c. Coccidioidomycosis
   d. Ringworm
   e. Histoplasmosis
4. Which of the following is *not* a disease caused by a protozoan?
   a. Malaria
   b. Trichomoniasis
   c. Amebic dysentery
   *d. Encephalitis
   e. African sleeping sickness
5. Which of the following will *not* take a Gram stain so must be acid-fast stained?
   a. *Staphylococcus aureus*
   b. *Pseudomonas*

*c. *Mycobacterium tuberculosis*
   e. *Escherichia coli*
   e. *Streptococcus pyogenes*
6. Which one of the following techniques would you use to determine the number of *viable* microorganisms per milliliter of broth?
   a. Turbidity tests
   *b. Plate count following dilutions
   c. Slide micrometer
   d. Spectrophotometer
7. A gram-positive bacterium is
   a. red
   *b. purple
   c. green
   d. colorless
   e. none of the above
8. A gram-negative bacterium
   a. is decolorized by alcohol
   b. takes the counterstain
   c. stains red with safranin
   *d. all of the above
   e. none of the above
9. A culture that contains only one species or type of bacteria is a
   *a. pure culture
   b. mixed culture
   c. contaminated culture
   d. all of the above
   e. none of the above
10. Streak-and-pour plates using nutrient agar are useful for
   a. studying biochemical differences
   *b. isolating bacterial colonies for pure cultures
   c. studying bacterial motility
   d. all of the above
   e. none of the above
11. Identification of a bacterial species depends on
   a. oxygen requirements
   b. nutrient requirements
   c. biochemical characteristics
   d. staining characteristics

*e. all of the above
12. Cocci that divide in two or three planes to form irregular clusters of cells are called
    a. diplococci
    b. streptococci
    c. sarcinae
    *d. staphylococci
    e. gaffkya
13. The spiral forms of bacteria include
    a. vibrios
    b. spirilla
    c. spirochetes
    *d. *b* and *c*
    e. none of the above
14. The common method of reproduction in bacteria is
    a. spore formation
    b. mitosis
    *c. binary fission
    d. all of the above
    e. none of the above
15. Structures that provide for motility in bacteria are
    *a. flagella
    b. cilia
    c. pseudopodia
    d. all of the above
    e. none of the above
16. Filamentous appendages, shorter and more numerous than flagella, observed on certain gram-negative bacteria are
    a. cilia
    *b. pili
    c. pseudopods
    d. all of the above
    e. none of the above
17. Which of the following men have been associated with the development of penicillin as an antibiotic?
    *a. Fleming
    b. Domagk
    c. Lister
    d. Waksman
    e. Koch
18. The word used for an agent that will inhibit bacteria is
    a. bactericidal
    *b. bacteriostatic
    c. fungicidal
    d. allergic
    e. homeostatic
19. Sterilization is synonymous with
    a. disinfection
    b. sanitization
    c. bacteriostatic
    *d. none of these
20. Which of the following materials is necessary for the serologic diagnosis of

typhoid fever using the serum of an infected person?
    *a. Typhoid exotoxin
    b. Fomite
    c. Differential media
    d. Typhoid organisms
21. Which of the following diseases is *not* caused by bacteria?
    a. Gonorrhea
    *b. Malaria
    c. Typhoid fever
    d. Diphtheria
    e. Tetanus
22. The spreading factor produced by certain streptococci is called
    a. fibrinolysin
    b. coagulase
    *c. hyaluronidase
    d. toxemia
    e. antigen
23. Which of the following substances aid infectious bacteria by spreading through tissue?
    a. Leukocidins
    b. Complement
    *c. Hyaluronidase
    d. Phagocytosis
    e. Coagulase
24. The simplest and most expedient means of *isolating* β-hemolytic streptococci from a sputum specimen is to
    *a. streak a plate of blood agar
    b. inoculate enriched broth
    c. inoculate a guinea pig
    d. prepare a gram stain
25. Inclusion bodies, such as Negri bodies, refer to
    a. bacteriophages within a bacterium
    b. spores within a bacterium
    c. reserve or stored food material
    *d. intracellular aggregates of viruses
26. The term *carrier* refers to
    a. an insect harboring a pathogenic organism
    b. an organism that has been made avirulent
    c. an organism infected with a bacteriophage
    *d. a person who harbors the causative organism of the disease, but who has no symptoms
    e. an organism capable of causing a disease
27. Contagious respiratory diseases may be transmitted from person to person by
    *a. droplets containing bacteria
    b. insect vectors
    c. food poisoning
    d. transfusions

e. water

28. The term *virulence* means
    a. toxigenicity
    *b. the ability to invade a host and produce disease
    c. aggressiveness
    d. antibody formation
    e. hemoglobinemia

29. Which of the following are *not* procaryotic "protista"?
    a. Cyanobacteria
    *b. Protozoa
    c. Bacteria

30. Rickettsias cause diseases in man, to whom they are transmitted by
    a. contact with rodents
    *b. bites of an arthropod vector
    c. contact with an infected human being
    d. eating infected food
    e. drinking soda water

31. A micrometer is
    *a. 1/1,000 of a millimeter
    b. 1/10,000 of a millimeter
    c. 1/100 of a millimeter

32. Organisms that live at the expense of other living beings are called
    a. saprophytes
    *b. parasites
    c. protophytes
    d. colonies
    e. leukocytes

33. Bacteria that grow best at temperatures between 15°C and 45°C are termed
    *a. mesophilic
    b. psychrophilic
    c. thermophilic
    d. aerobic

34. Bacterial populations are usually expressed as the number per
    a. loopful
    *b. ml (milliliter)
    c. tube
    d. slide

35. Which combination of data is required to plot the conventional growth curve of bacteria?
    *a. Log of number of living bacteria versus time
    b. Log of number of living and dead bacteria versus time
    c. Log of number of generations versus time
    d. Log of generation time versus time

36. A microorganism that grows in either the absence or the presence of free oxygen is known as an
    a. obligate aerobe
    *b. facultative anaerobe
    c. anaerobe

d. microaerophile

37. Lysis of red blood cells may be caused by
    a. complement
    b. a hemolysin
    c. a hypotonic solution
    *d. all of the above

38. *Staphylococcus aureus* has become resistant to which of the following?
    a. DDT
    b. Sodium pentothal
    *c. Penicillin

39. The yeastlike fungus *Candida albicans* can cause which of the following diseases?
    a. Rickets
    *b. Thrush
    c. Sickle cell anemia

40. Most skin cancers are caused by what environmental factor?
    a. Smog
    b. Wind
    *c. The sun's ultraviolet light radiation

41. The most effective way to disinfect drinking water at present is
    a. fluoridation
    *b. chlorination
    c. filtration

42. Which of the following men disproved by experimentation the theory of spontaneous generation?
    a. Leeuwenhoek
    b. Needham
    *c. Pasteur
    d. Koch

43. Lister's contribution to microbiology was associated with
    *a. antiseptic surgery
    b. the discovery of Listerine
    c. the design of the Lister bag
    d. the preservation of foods

44. The man given credit for the first description of bacteria was
    a. Pasteur
    *b. Leeuwenhoek
    c. Gram
    d. Koch

45. The mechanism for transfer of genetic material from one bacterial cell to another through actual contact is
    *a. conjugation
    b. transduction
    c. phage conversion
    d. transformation
    e. none of the above

46. Rickettsias are usually described as obligate parasites because
    *a. they grow and multiply only within living cells
    b. they can be cultivated on blood agar
    c. they cause disease in humans

d. all of the above
e. none of the above

47. Rickettsias may be grown in the laboratory by
    a. infecting susceptible animals
    b. inoculating tissue cultures
    c. inoculating chick embryos
    *d. all of the above
    e. none of the above

48. The smallest known organisms capable of growth and reproduction outside of living cells are
    a. rickettsias
    b. bacteria
    *c. mycoplasmas
    d. bdellovibrios
    e. none of the above

49. Why is marked pleomorphism characteristic of mycoplasmas?
    a. They possess a flexible wall external to their membrane.
    *b. They lack a rigid cell wall.
    c. Their shape changes as the cells age.
    d. All of the above.
    e. None of the above.

50. How can bacteria be induced to form L-forms?
    a. By the addition of toxic chemicals to a culture
    b. By the addition of an antibiotic which interferes with the synthesis of peptidoglycan
    c. By treatment of an infection with penicillin
    *d. All of the above
    e. None of the above

51. Viruses lack the following component(s), which is (are) essential to replication outside a cellular medium
    a. an outer protein boundary
    b. nucleic acid
    c. an ATP-generating system
    d. ribosomes for protein synthesis
    *e. only c and d

52. A virus that infects a bacterial cell is called
    *a. a bacteriophage
    b. an inclusion
    c. a chromatid
    d. all of the above
    e. none of the above

53. When a bacterial culture is infected with a temperate phage, the culture is said to be
    a. temperate
    *b. lysogenic
    c. stationary
    d. all of the above
    e. none of the above

54. To produce exotoxin, a culture of *Corynebacterium diphtheriae* must be

a. temperate
*b. lysogenic
c. stationary
d. all of the above
e. none of the above

55. Negri bodies are useful in the diagnosis of
    *a. rabies
    b. measles
    c. smallpox
    d. vaccinia
    e. none of the above

56. In comparison with bacteria, true fungi are
    a. larger and more complex
    b. eucaryotic
    c. unicellular and multicellular
    *d. all of the above
    e. only a and b

57. The most common method of reproduction in molds is
    *a. spore formation
    b. fission
    c. budding
    d. all of the above
    e. none of the above

58. The fungi that cause infections of the hair, skin, and nails are known as
    a. systemic fungi
    b. actinomycetes
    *c. dermatophytes
    d. all of the above
    e. none of the above

59. Lipids are commonly found, along with proteins as part of
    *a. plasma membranes
    b. cell walls
    c. ribosomes
    d. nuclei
    e. solutions

60. Lipid molecules are composed of
    a. amino acids
    *b. glycerol and fatty acids
    c. sugars
    d. nucleotides
    e. only carbon atoms

61. Carbohydrate molecules include
    a. glucose
    b. starch
    c. cellulose
    d. lactose
    *e. all of the above

62. The order, or sequence, of amino acids in a protein molecule constitutes its
    *a. primary structure
    b. secondary structure
    c. tertiary structure
    d. active site
    e. specificity

63. Amino acids are the subunits of which compound?

a. Carbohydrates
b. Fats
c. Phospholipids
*d. Proteins
e. Nucleic acids

64. Proteins that speed up the rate of biological reaction are called
  a. proteases
  b. transferases
  *c. enzymes
  d. mucopolysaccharides
  e. phospholipids

65. Proteins are polymers of
  a. sugars
  b. enzymes
  *c. amino acids
  d. glycerol and fatty acids
  e. nucleotides

66. DNA is a polymer of
  a. deoxyribose
  b. phosphate
  c. organic bases
  *d. nucleotides
  e. peptides

67. Polysaccharides are polymers of
  a. nucleotides
  b. amino acids
  *c. glucose
  d. fatty acids
  e. sucrose

68. All of the following characterize DNA *except* that
  *a. it is replicated on the ribosome
  b. its shape is described as a double helix
  c. it contains the cell's protein code
  d. it may mutate
  e. it can produce exact copies of itself

69. Translation is best described as the direction of
  a. mRNA synthesis by DNA
  b. tRNA synthesis by DNA
  *c. protein synthesis by mRNA
  e. DNA synthesis by mRNA
  e. mRNA synthesis by tRNA

70. Dermatomycosis refers to
  a. viral eye disease
  b. bacterial skin infection
  *c. fungal skin infection
  d. protozoan skin infection
  e. middle layer of the skin

71. Salmonella infection is spread
  a. by food handlers
  b. contaminated water
  c. contaminated food
  d. *a* and *c* only
  *e. all of the above

72. Bacterial endocarditis is a disease of the
  a. lymphatic system
  *b. blood-vascular system

c. nervous system
d. liver
e. salivary glands

73. Suppose you were a pharmacist and you received a prescription for chloraquine. You would surmise that the individual who presented the prescription had
  *a. malaria
  b. plague
  c. toxoplasmosis
  d. trachoma
  e. rheumatic heart disease

74. Meningitis is caused by
  a. *Neisseria meningitidis*
  b. *Cryptococcus neoformans*
  c. *Streptococcus pneumoniae*
  d. two of these
  *e. all of these

75. Which of the following diseases is (are) caused by viruses?
  a. Mumps
  b. Poliomyelitis
  c. Aseptic meningitis
  d. Two of these
  *e. All of these

76. Peritonitis is a disease of the
  a. liver
  b. salivary glands
  c. lymphatic system
  d. nervous system
  *e. none of these

77. The bacteria found in dental plaque are chiefly
  *a. acid-producing
  b. endotoxin-producing
  c. proteolytic
  d. two of these
  e. all of these

78. Which one of the following diseases demonstrates synergism?
  a. Typhoid fever
  b. Actinomycosis
  c. Candidiasis
  *d. Dental caries
  e. Cholera

79. We associate the letters DPT with which one of the following diseases?
  a. Streptococcal pneumonia
  b. Mycoplasma pneumonia
  c. Tuberculosis
  d. Scarlet fever
  *e. Diphtheria

80. Croup is a form of
  a. pneumonia
  b. whooping cough
  c. influenza
  *d. the common cold

81. Gingivitis occurs in the
  *a. oral cavity

b. pharynx
c. stomach
d. small intestine
e. large intestine

82. The toxin that causes botulism is a (an)
   a. endotoxin
   b. enterotoxin
   c. neurotoxin
   *d. two of these
   e. all of these

83. "Rice water" stools are particularly characteristic of
   a. typhoid fever
   b. shigellosis
   *c. cholera
   d. vibriosis
   e. amebiasis

84. Which one of the following diseases is most likely to result from autoinfection?
   a. Chickenpox
   b. Rubella
   c. Rubeola
   d. Tetanus
   *e. Impetigo

85. The causative agents of tetanus, anthrax, and gas gangrene are all
   a. endospore formers
   b. bacteria
   c. anaerobes
   *d. two of these
   e. all of these

86. *Giardia lamblia* protozoa
   a. have easy-to-recognize trophozoites because they have bilateral symmetry
   b. have worldwide distribution
   c. are most common in cold water
   *d. all of the above
   e. *a and b*

87. *Trichomonas vaginalis* is characterized by
   a. one anterior flagella
   b. four posterior flagella
   *c. undulating membrane
   d. all of the above

88. *Pneumocystis carinii*
   a. causes pneumonia only in immunocompromised individuals
   b. exists worldwide in the digestive tracts of a variety of animals
   c. produces infections that are associated with AIDS
   d. all of the above
   *e. *a and c*

89. Sources of clostridial spores are
   a. soil, especially farm land
   b. normal flora of the intestine
   c. dust
   d. rivers
   *e. all of the above

90. An animal suspected of being rabid should be quarantined to
   a. test for Negri bodies should the animal die
   b. isolate the animal to prevent further spread of infection
   c. observe the animal for signs or symptoms of rabies
   *d. all of the above

Matching

Match the causative agent with the disease.
A. Virus
B. Rickettsias
C. Bacteria
D. Fungi
E. None of the above

(A) ___ 91. Rabies
(C) ___ 92. "Strep" sore throat
(D) ___ 93. Thrush (candidiasis)
(C) ___ 94. Diphtheria
(E) ___ 95. Giardiasis
(C) ___ 96. Tetanus
(B) ___ 97. Rocky Mountain spotted fever
(A) ___ 98. Poliomyelitis
(E) ___ 99. Amebiasis
(C) ___ 100. Botulism
(B) ___ 101. Typhus fever
(C) ___ 102. Salmonellosis
(A) ___ 103. Hepatitis
(D) ___ 104. Ringworm
(A) ___ 105. Smallpox

True or false

(F) ___ 106. In Gram staining, the gram-negative organism retains the color of the first dye after the decolorizer is used.

(T) ___ 107. The first step in isolating an organism in a pure culture is the streak plate technique.

(T) ___ 108. Gram-positive or gram-negative information may help the doctor determine which drugs would be useful in therapy.

(F) ___ 109. Acid-fast organisms include the pathogens *Mycobacterium tuberculosis* and *Streptococcus pneumoniae.*

(T) ___ 110. The phenol coefficient indicates the effectiveness of a disinfectant as compared with that of phenol

(T) ___ 111. Complement is normal serum